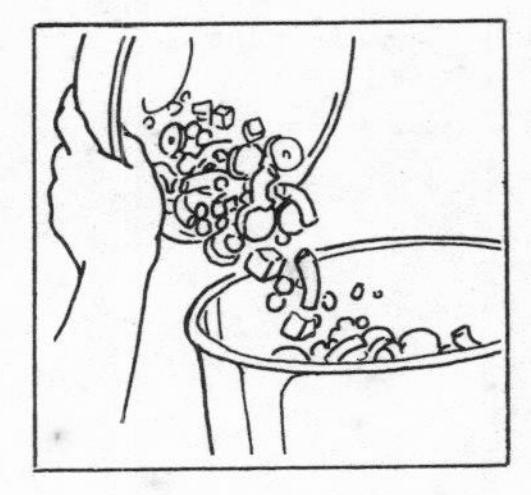
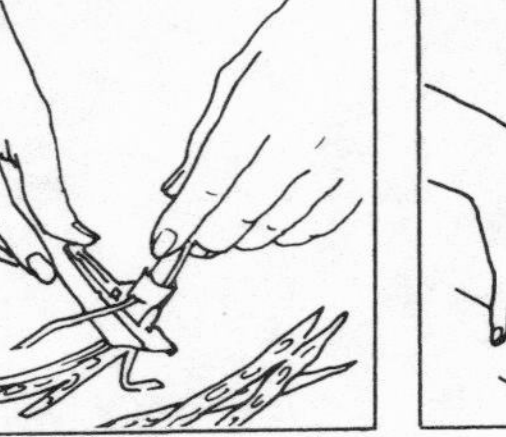
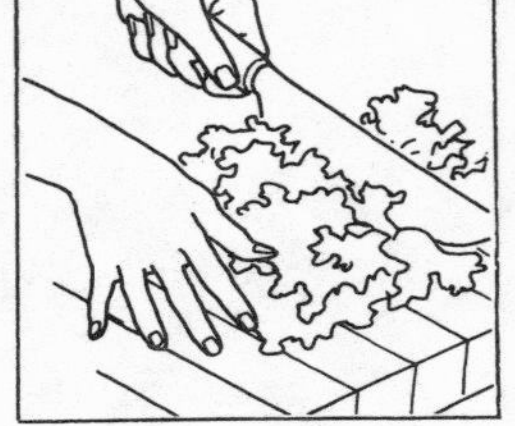
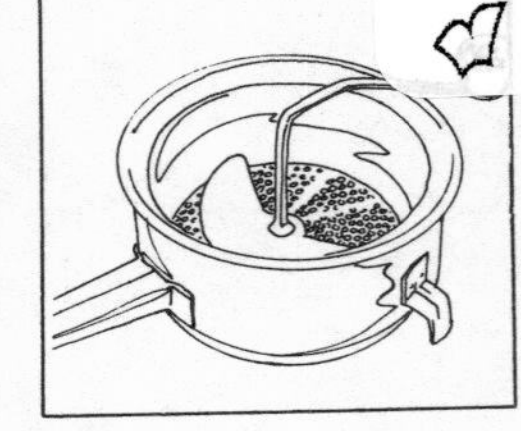

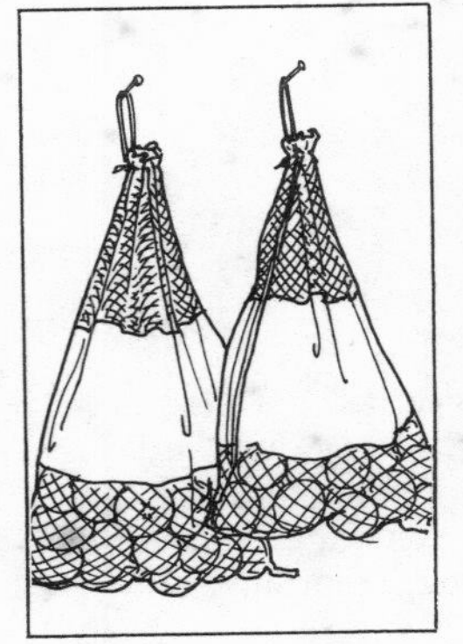

The *Busy Person's Guide to Preserving Food*

Janet Bachand Chadwick

GARDEN WAY PUBLISHING CHARLOTTE, VERMONT 05445

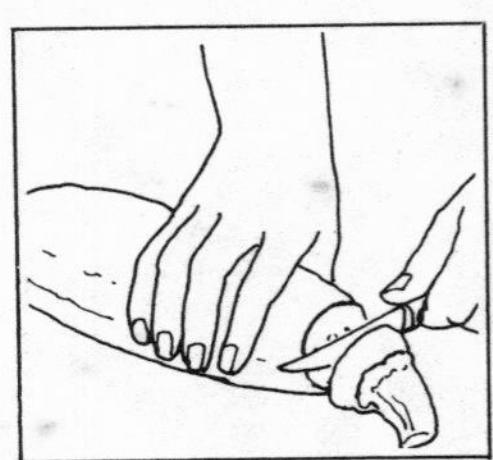
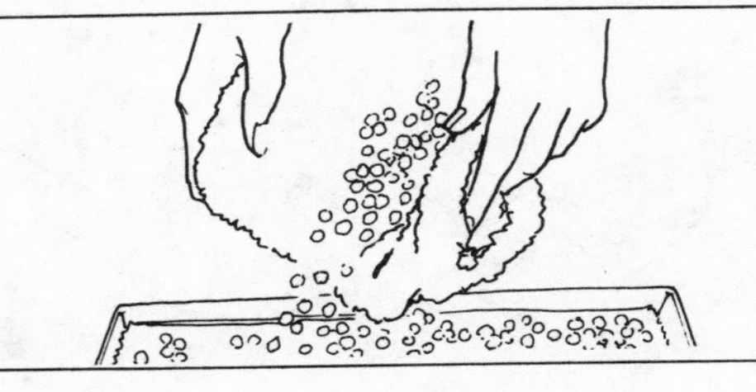

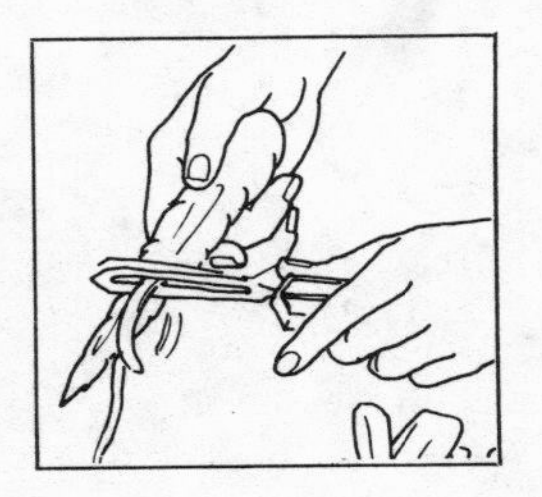

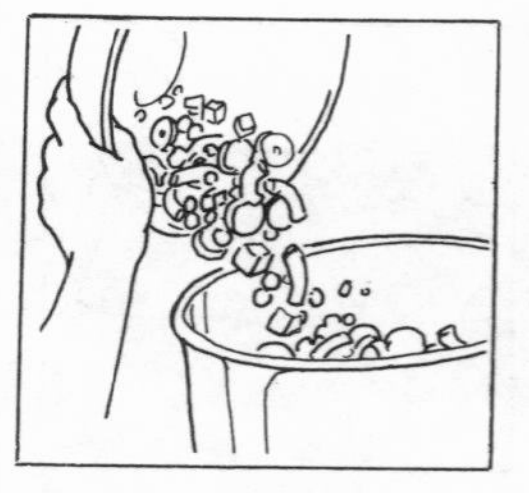

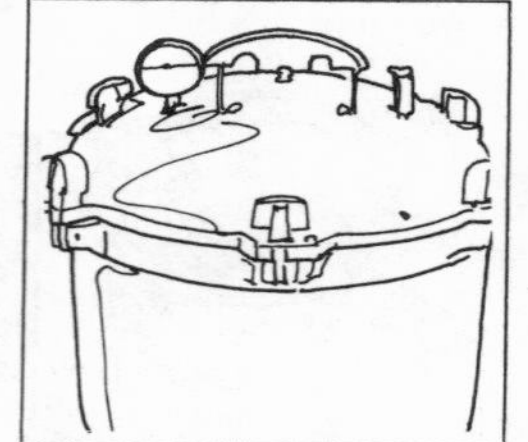

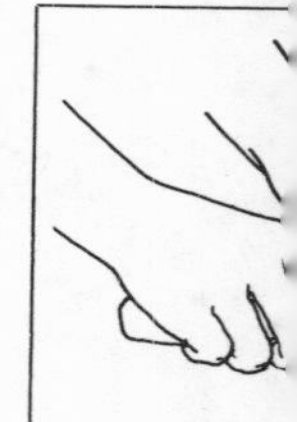

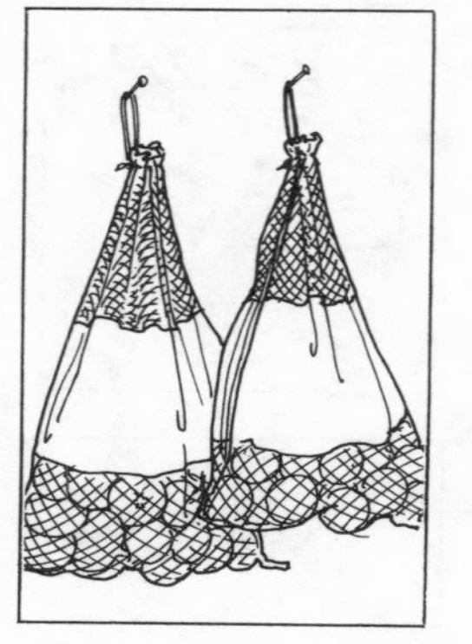

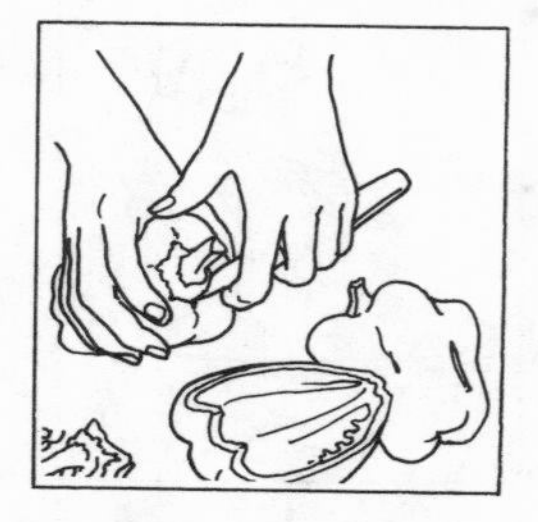

Illustrations and cover art by Elayne Sears

Library of Congress Cataloging in Publication Data

Chadwick, Janet, 1933–
The busy person's guide to preserving food.

Bibliography: p.
Includes index.
1. Food—Preservation. I. Title.
TX601.C43 641.4 82-1022
ISBN 0-88266-263-5 AACR2

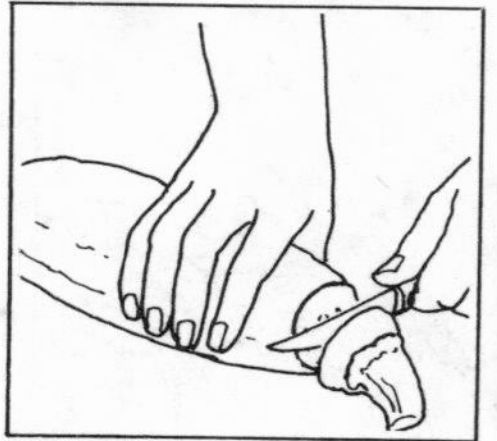

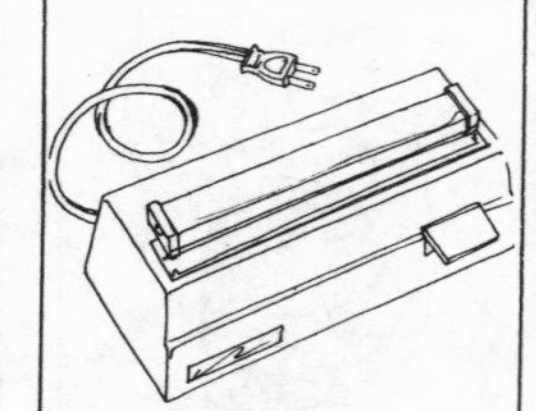

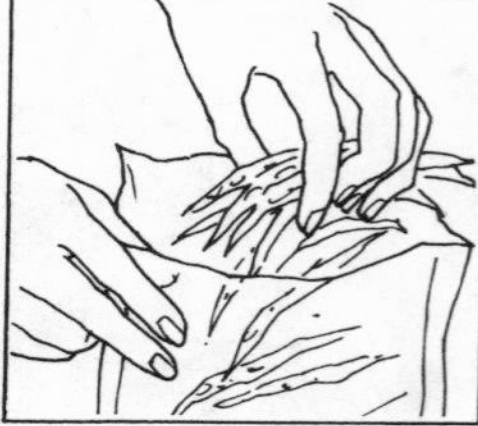

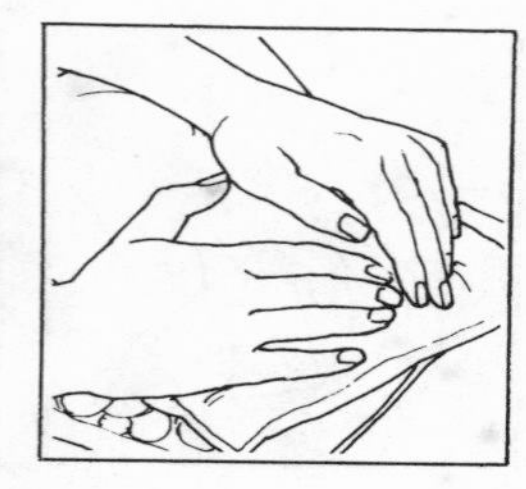

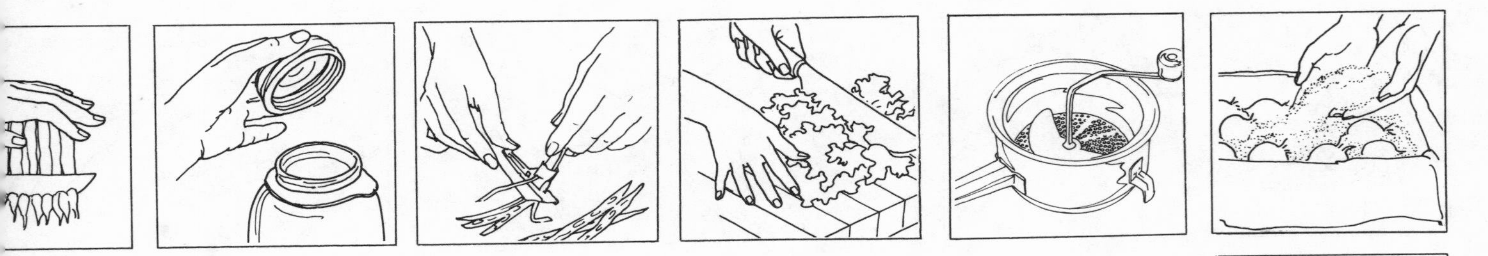

To my mother, Florence Bachand, and my grandmothers, Emma Bachand and Angeline Guay

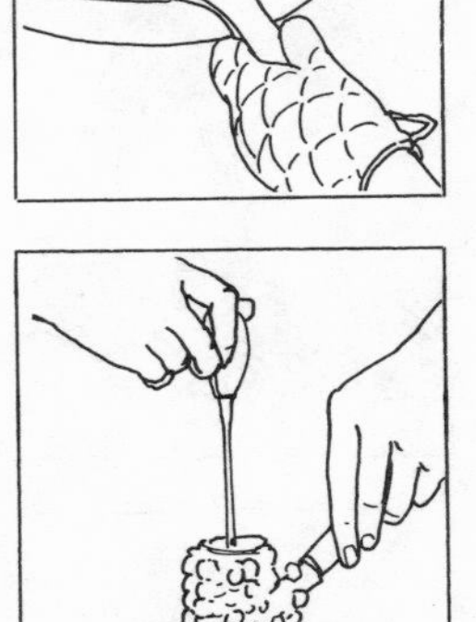

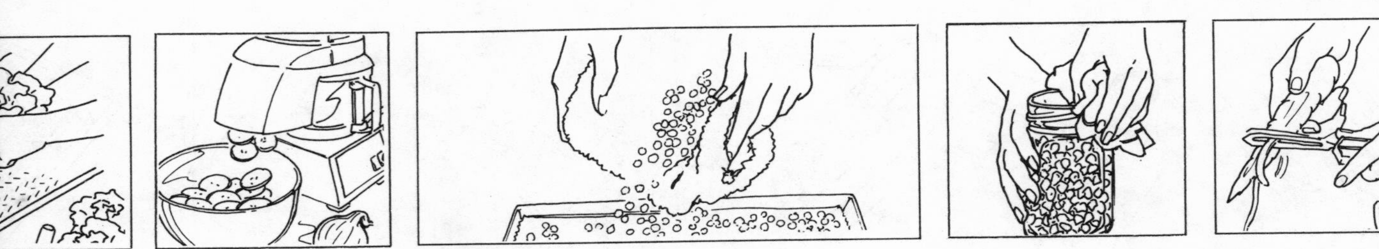

PEAS
PEAS

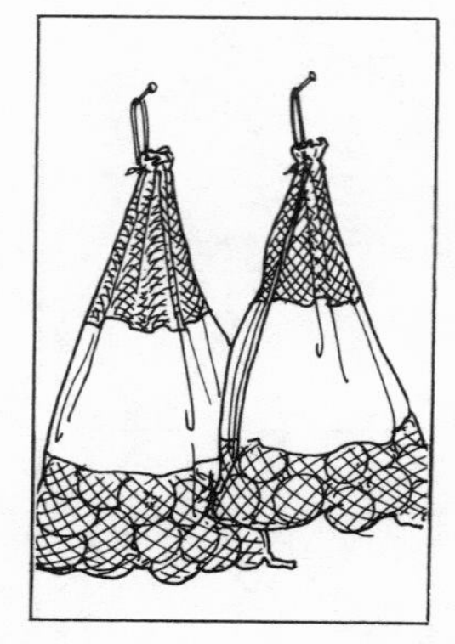

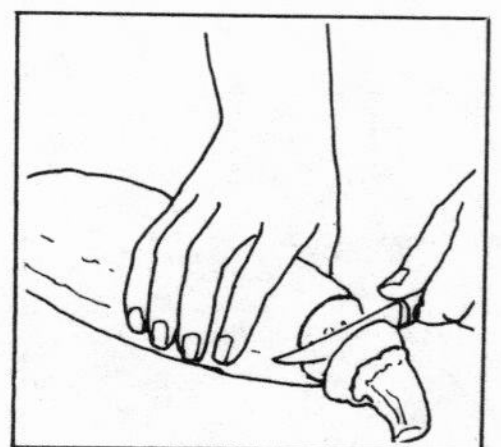

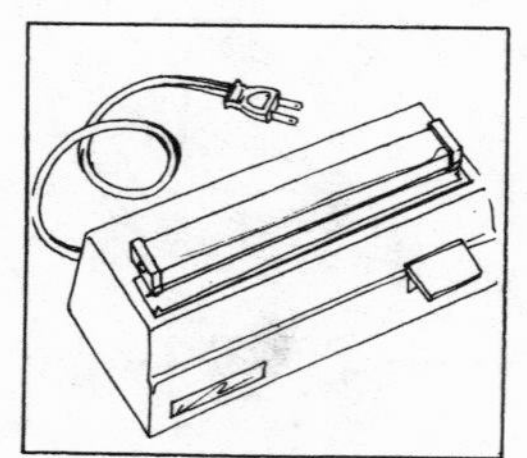

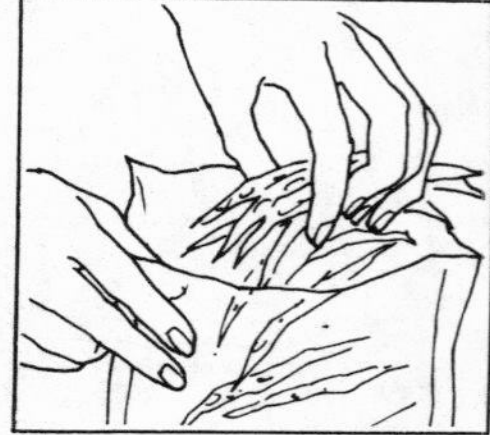

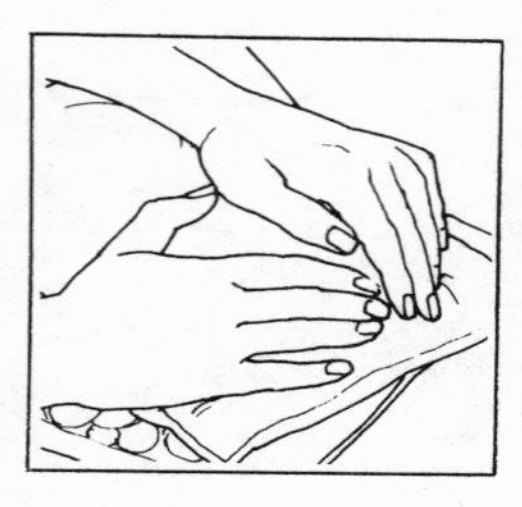

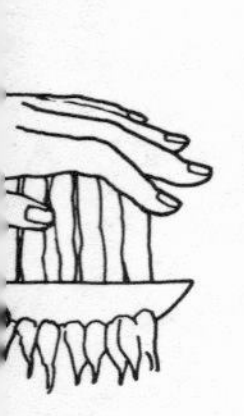
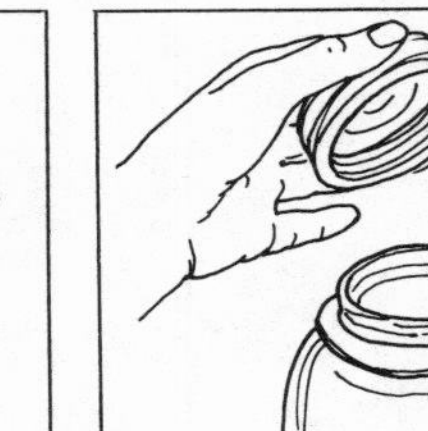
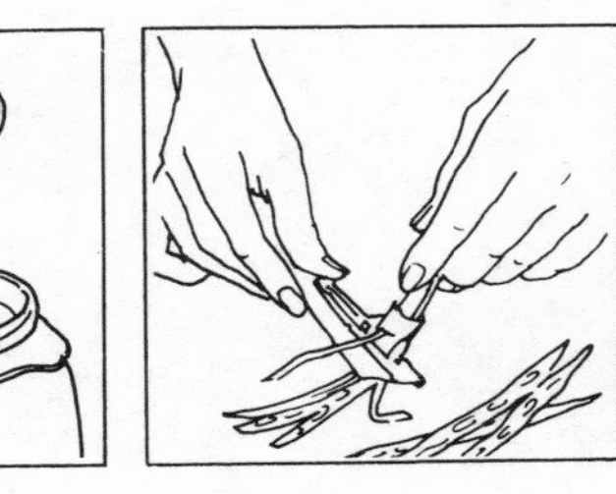
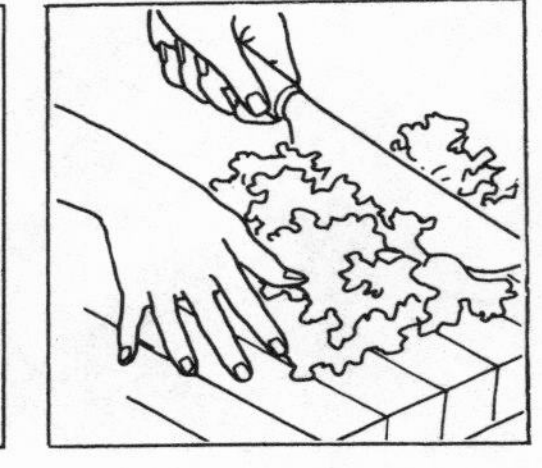
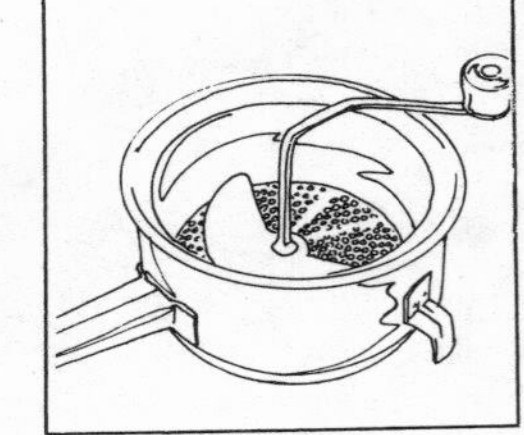
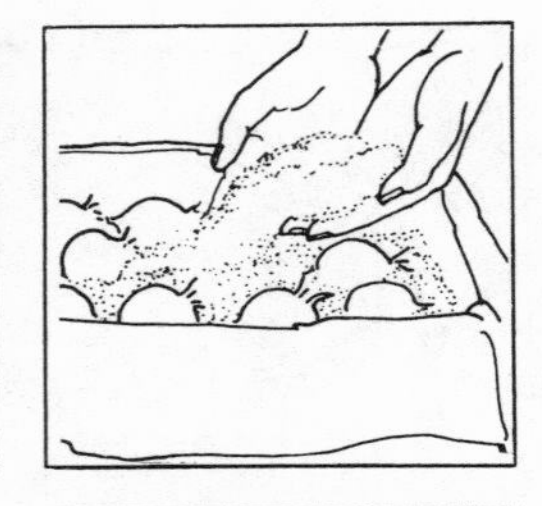

Contents

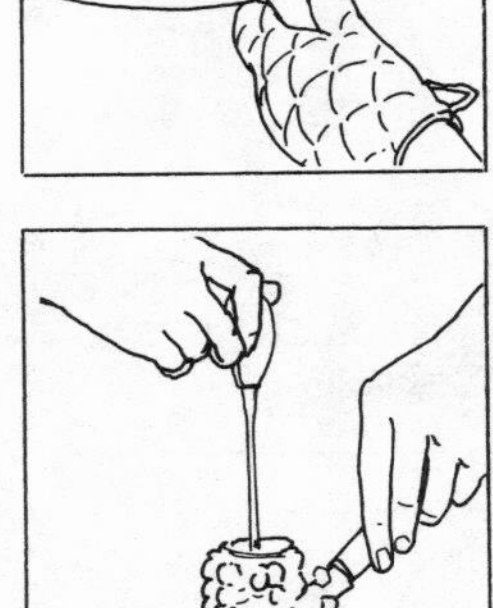

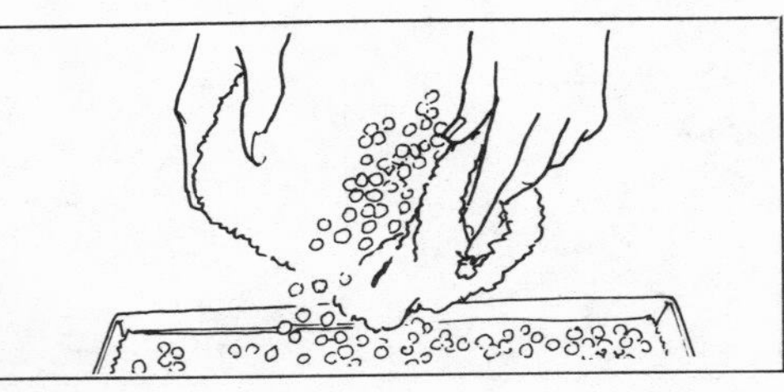
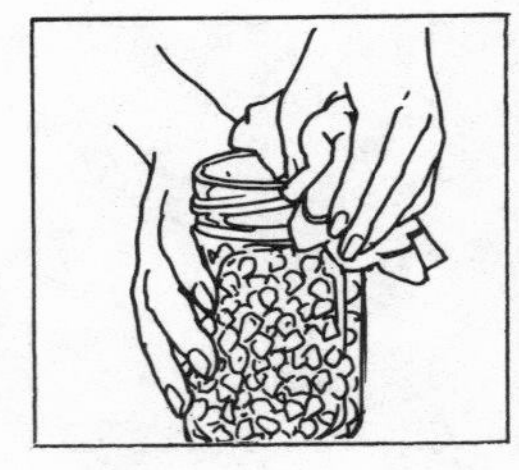
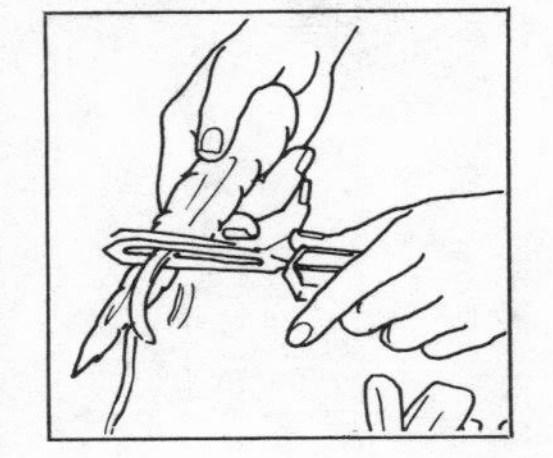

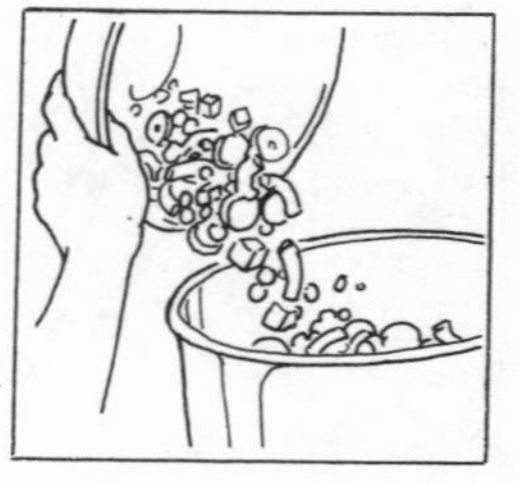
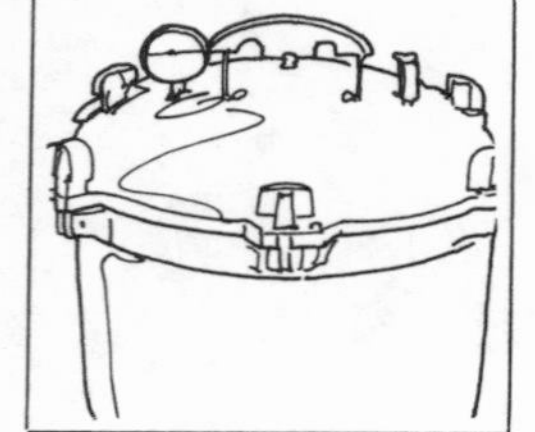
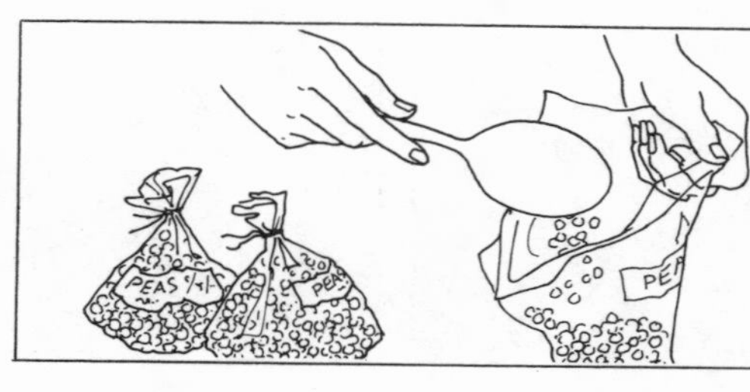
PEAS

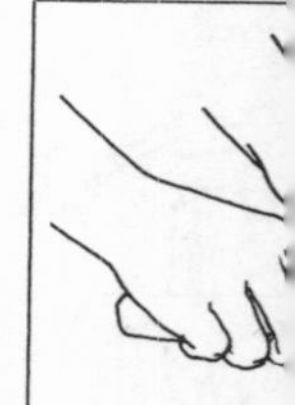
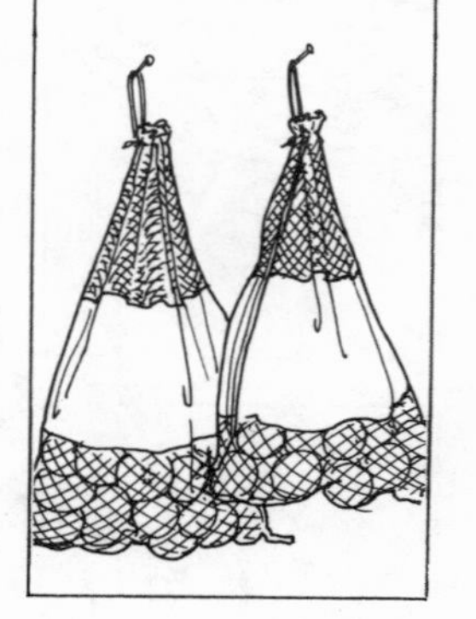

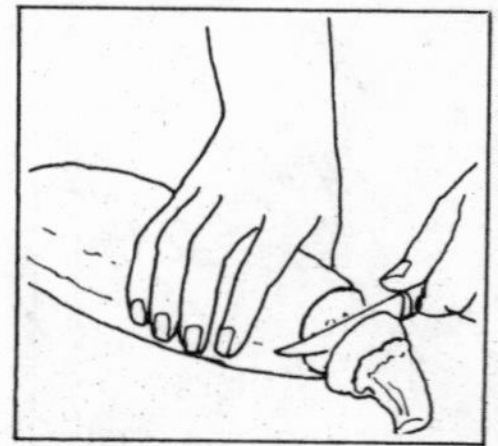
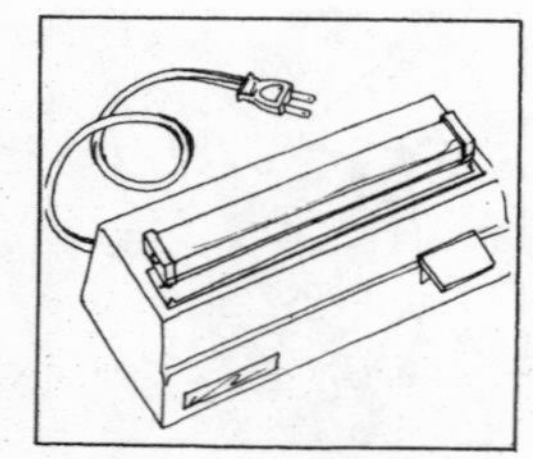

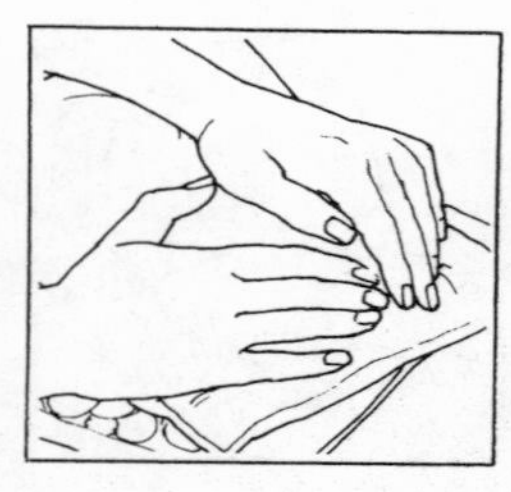

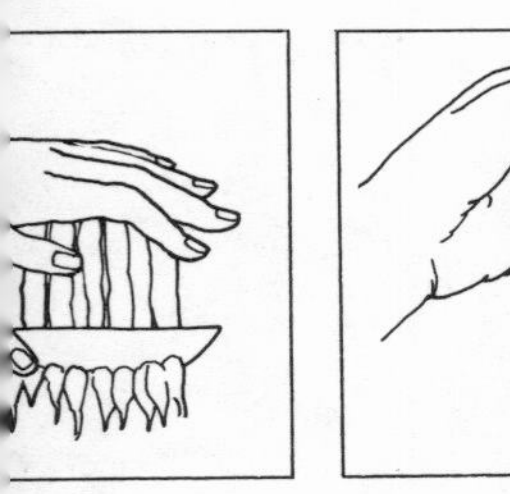
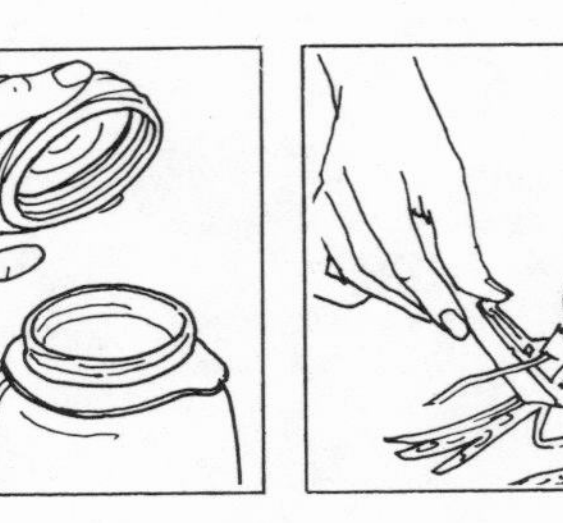
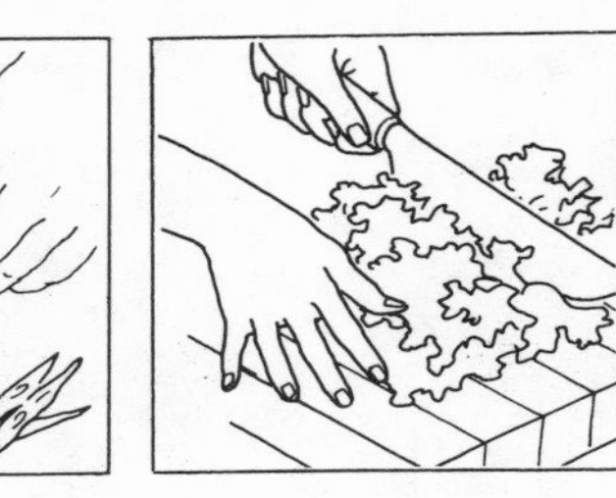
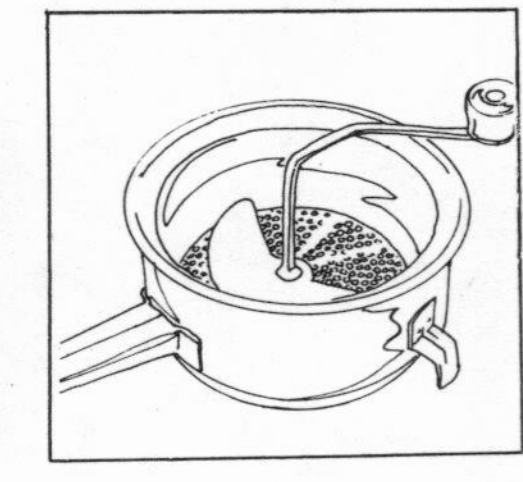
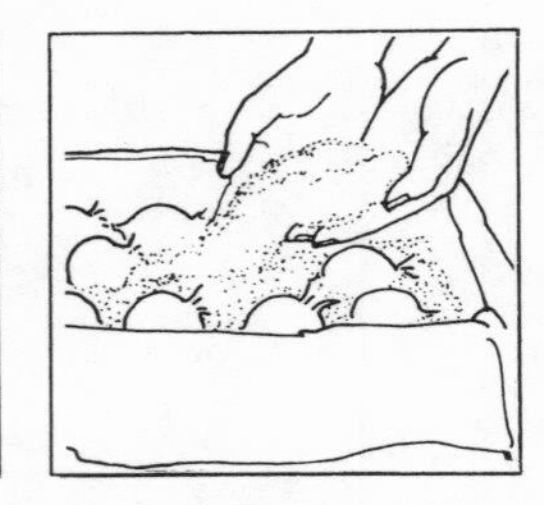

Acknowledgements

I would like to thank Louellan Wasson, home economist, for her technical editing and moral support throughout the writing of this book. Many others deserve thanks, too: Andrea Chesman, my editor, who had the original idea for this book; Mary Clark of Garden Way, for obtaining many new items of food processing equipment for me to test; Moulinex and General Electric companies, in particular Mr. Ted Miller of General Electric Co., for making available to me food processors for testing; Lynn Liberty of Garden Way Living Center, for her never-failing support; Win Way, University of Vermont Extension Agronomist, for his educated palate; and Elayne Sears for her descriptive illustrations.

Last but not least, I'd like to thank my family, especially my husband Raymond, for taking on extra gardening chores to help make this book possible, and to my daughter Kim, for her all-around help.

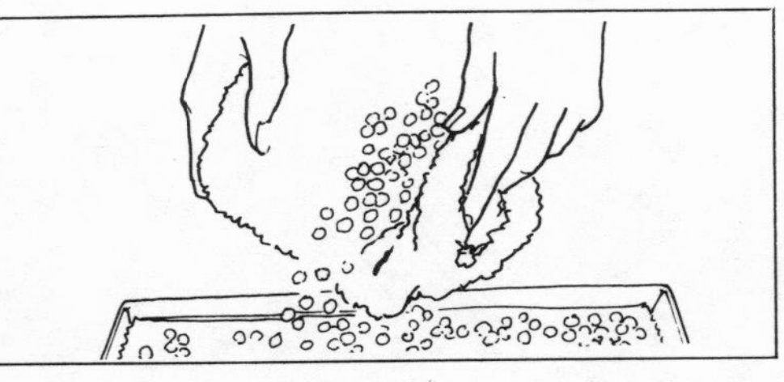
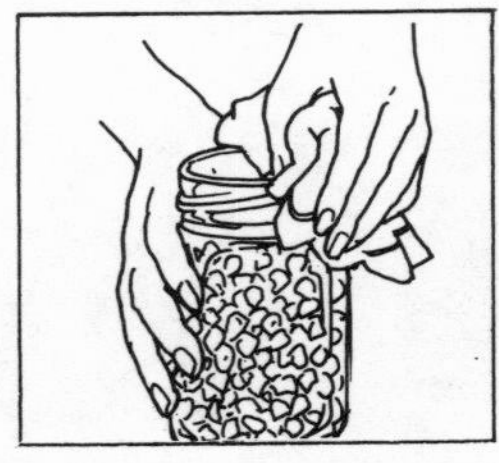
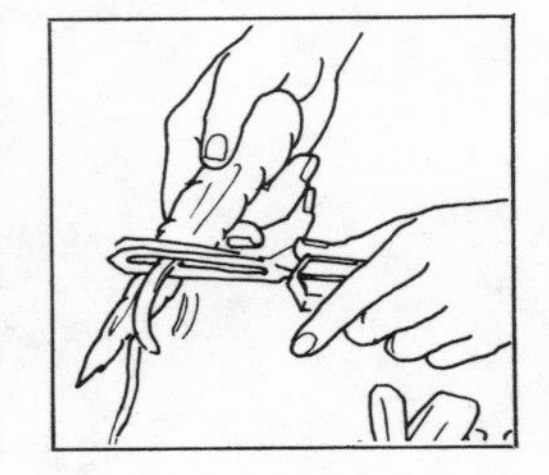

Introduction

You have three children under five, the baby cried all night, and your mother-in-law is coming for supper. What are you going to do with that bushel of green beans? . . . Or, it's Thursday, and you just arrived home after a hard day at the office. You know that if you wait until Saturday to make dill pickles, the cucumbers will be too large; but if you pick them now, they will be punky by Saturday . . . but you don't feel like pickling until midnight. What are you going to do? Well, take a deep breath and relax, *this* book is for you.

Vegetable gardening is rapidly becoming the number one American pastime. Escalating food costs, inferior quality products, and chemical additives have convinced many of us that the best way to provide good food at reasonable costs for our families is to raise it ourselves.

When calculating the savings of a garden and home food processing, many people will remark (always non-gardeners), "Yes, but how much is your time worth?" To answer that question you have to be honest with yourself. Would you really be holding down an extra job during those hours that would *net* you more cash in the bank? Can you put a price tag on the physical, emotional, and spiritual satisfactions that you derive from working the soil and producing the foods that nourish your family? Is the kind of food you feed your family important to you? Is there any way you can "buy" the feeling of pride you have when you show off the full freezer; the rows of canned vegetables, fruits, pickles, jams, and jellies; or the root cellar shelves filled to the ceiling? If these things are not important to you, then maybe you should not be gardening or preserving food; because there *is* a lot of work involved—but the work should be a joy and a challenge to your ability to be more self-reliant.

I've written this book for those of you who garden and hold down outside jobs, or are busy with youngsters and other outside activities. I know that it is hard enough to cope with the normal routine of family, home, and job, along with raising a small garden for fresh summer vegetables, without trying to squeeze in long hours of food preservation at the end of a busy day. But much food can be preserved for storage in the small blocks of time you have available on a daily or weekly basis. This book can be used as a primer if you are new to the art of food preservation, and it may bring new and exciting ideas to those of you who have been preserving food for years.

How Can This Book Help You?

Freezing is the most popular method of food preservation. It's the most time-saving method, and it produces the best finished product; and so this book will focus strongly on freezing techniques, including some new methods that are even more time-saving and produce an even better finished product.

I will discuss the equipment that is absolutely necessary for food preservation, and I will add information on other equipment that, while not absolutely necessary, will make food preservation faster, easier, and, in general, give a better finished product.

People who are serious about food preservation should realize that many steps can be taken year-round to make harvest time easier. Throughout the book I will give you tips that will help you to make the best use of weekends, weekdays, and overnight hours. These tips will include the best times to harvest vegetables as well as the best ways to keep them fresh for up to three days, so that you can plan to harvest one day and preserve a day or so later.

Since I can already hear nutritionists raising their voices in a collective uproar over this last remark, I think I'd better explain myself. Many commercially sold fresh vegetables and fruits are treated with chemicals to preserve their *look* of freshness; they can be from two days to two weeks old when they appear in your market. You have no way of knowing how many times frozen foods have defrosted and have been refrozen (losing nutrients and flavor) before they appear in the frozen food cases of your market. Nor is there any way to judge the hours or days that lapsed between the harvest and processing of canned foods. You will never know what chemicals were sprayed on these foods while they were growing.

Obviously, you can't possibly get anything worse than what you buy in a supermarket, so you are already ahead of the game. Even though the *best* way to preserve all the nutrients and fresh flavor of food is to pick them at their peak and process these foods immediately, you can be sure that waiting even two days between harvest and processing will not cost you much more than minimal loss of either nutrients or flavor *if you follow my instructions on storing*. Therefore, you have nothing to lose and everything to gain by preserving your food by my methods.

I will give step-by-step, illustrated instructions for processing each commonly grown vegetable by the simplest method available. These methods will all be timed to give you an idea of approximately how long it will take to accomplish the tasks at hand.

I will include food preparation ideas and recipes for delicious meals that can be prepared ahead of harvest time to make busy days much easier. I will also share my recipes that incorporate the vegetables being preserved for quick and easy harvest meals.

In addition to all this, I will include a list of reference books on gardening and food preservation that can be read during leisure hours. (Winter is a good time for this.)

There are three important pieces of advice I'd like to share before we get into serious food preservation. First, take the time to excite the entire family with the real meaning of what you are accomplishing for yourselves, and how their independence will make them stand taller in a world that is so filled with deprivation. It will make the harder days easier for everyone to cope with. A good attitude is insurance against discouragement when you are tired. Second, stay flexible. This is important mostly from a psychological standpoint. If you become too rigid in your expectations, you will waste tremendous amounts of emotional energy trying to keep up, or in disappointment at failures. This can ruin the best of goals. Finally, know yourself. If you work best at night, plan your work for the evening hours; but if you are a morning person, get up an hour or so earlier in the morning to accomplish extra tasks.

The following pages are packed with information to help you make the best use of your time and resources during the harvest season. Good luck and have fun!

CHAPTER 1

Taking the Mystery Out of Equipment

Pea shellers, apple peelers, cherry pitters, food mills, food processors: Are they really time-savers in the harvest kitchen?

If you are serious about making food preservation as quick and easy as possible, you will want to invest in some time-saving equipment. The right equipment not only makes work easier and more enjoyable, but the end results are usually better, so the feeling of satisfaction is greater.

You do not need to purchase the most expensive equipment on the market; there are some very good and reliable products sold at reasonable prices. A good food processor is just about the best helping hand a busy person can have at harvest time. Prices range from as low as $45, all the way up to $200 for one of the best Cuisinart models with all the attachments. Once you realize what a help this type of appliance will be to you throughout the year, you will probably see it as a good investment. On the other hand, every new gadget that you see advertised may not be worth the money. In this chapter I will concentrate on the food-processing items that I have tested and found useful in my kitchen.

Determine Your Needs First

Before you step into a store, or look through a catalog, you should have a good idea of just what equipment you need. What you need is determined by the foods you will prepare and the methods of preservation you will use. Will you freeze? Can? Pickle? Store in a root cellar? Will you need to chop? Slice? Dice? Grind? Or puree?

If you haven't decided just how you will be preserving your harvest, glance ahead to chapters 3 and 4 to decide which methods you will be trying. There is no point in investing in equipment you won't use.

Basic Equipment List

For Root Cellaring: The Fastest Method

boxes
barrels
large plastic bags
crocks

For Freezing: The Next Fastest Method

freezer
scrub brush
blanching kettle
strainers—fine sieves and colanders
paring and chopping knives
food processing equipment—grinders, slicers, blenders, or food processors
measuring cups and spoons
wide-mouth funnel
timer
hot pads, mitts, or heavy potholders
heavy bath towels
large bowls
large roaster or kettle
cookie sheets or jelly roll pans for tray freezing

For Canning, Add to the Freezer List
canner—steam, boiling water bath, or pressure
preserving kettle
tea kettle
nonmetallic spatula or wooden chopstick
large wooden and slotted spoons
soup ladle
jar lifter or tongs
canning jars and lids

The Expensive Appliances

In my kitchen, I have found that I really save time by using my freezer, dishwasher, and food processor. Of course, you can preserve food without these appliances, but their year-round uses justify their costs.

Freezers

Although a freezer is a major investment, freezing is very often the fastest method of food processing, and it generally gives you a product that is closest to fresh.

Small freezers contained in refrigerators are not cold enough to freeze and store foods. Your freezer must be a separate unit that maintains a temperature of at least 0 degrees F. or lower to inhibit the growth of bacteria, yeasts, and molds, and to stop the enzyme action that can cause discoloration and destroy the fresh flavor of frozen foods.

Costs. You can buy a good freezer from a large chain store, such as Sears or Montgomery Ward, at prices as much as a third lower than some of the brand name models. There are operating costs associated with freezers which you should take into consideration. New energy-efficient models cost about $50 a year to operate. Older models may cost about $20 per year more, a consideration if you are contemplating a used freezer.

To the cost of energy required to run the freezer, add the energy it takes to process the food—about $10 (enough food for a family of 4). On top of that, add the cost of packaging frozen foods—about $50.

For Energy Efficiency, It Pays to Shop Around. Each freezer should list the approximate energy consumption per year of the unit, so be sure to do some comparison shopping.

In the January 1978 issue of *Consumer Report* magazine, the editors stated that the self-defrosting freezers may not cost more to operate than manual-defrosting models. Accumulated frost in a manual defrosting model acts as an insulator between the cooling coils and the interior, so that additional energy is required to keep the unit cold and to recool the entire freezer after it has been defrosted.

Chest Freezers Versus Upright Freezers. Chest freezers tend to be colder since cold air does not rise quickly when the lid is raised. They hold more food overall per cubic foot and, usually cost less than upright models.

Upright freezers use less floor space, and they make reaching for food more convenient. But cold air spills from them when the doors are opened; therefore, operating costs are greater. Much of the space in an upright goes to waste since it must be packed so that food won't fall out when the door is opened.

Food Processors Save Time

Food processors work so fast that washing the vegetables, cutting them to size, and placing them in the feed tube takes more time than the actual processing. This versatile appliance combines many of the functions of the blender, grinder, and slicer. While processors do not puree as efficiently as blenders, they do a creditable job. They will also grind coffee and meat, and slice, julienne, shred, and grate. Many have accessories that will allow you to blend, ripple cut, and cut french fries. They make breads, pie crusts, and sauces with just a flick of the wrist. They can be taken apart easily for cleaning, and most parts are dishwasher safe.

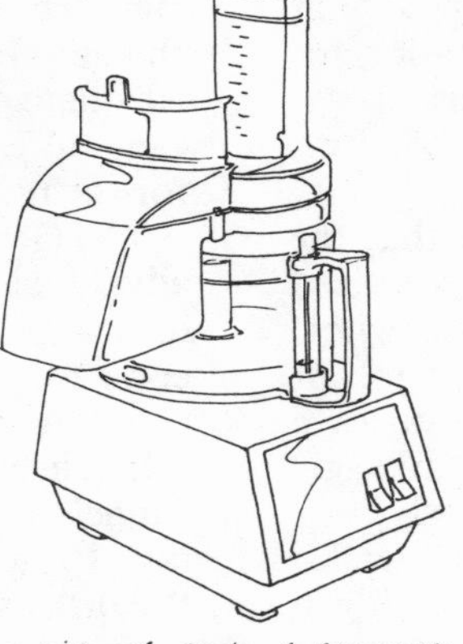

side chute food processor

Which Food Processor to Buy. In the September 1981 issue of *Consumer Reports* magazine, there is a report on tests of a wide range of processors. What this report did not cover was how well the processors performed on specific food preservation chores; so I tested several different brands of processors in my kitchen to find the processor that works the best for the harvest season.

There are 2 basic types of processors: direct-drive, which means that the bowl and blades sit directly on top of the motor's shaft; and belt-driven, where the bowl mounts to one side of the motor. The belt-driven types take up more counter space than the direct-drive processors, and the belt-driven machines that I have tested seem to overheat faster than those with direct-drive.

Most food processors feed the chopped, diced, or sliced vegetables into a bowl. The larger the bowl, the fewer times you will have to empty the vegetables into a larger bowl, and the less time your work will take. Some processors have a side-discharge chute, which sends the processed foods directly out of the processor and into whatever container you want, allowing you to process large quantities without stopping to empty the bowl and reassemble the unit.

Here is a look at the machines I tested. While I do not mean to disparage any manufacturer, I do find differences in how well the machines perform, and I think my findings will be of value to you.

Cuisinart and Robot-Coupe have a variety of models

that sell in the $100–200 range and are both very similar in design and quality. Cuisinart has the reputation for being the best-seller; but that is largely because it was the first food processor to be marketed in this country, and therefore is the most widely known.

The main advantages to these brands are that the most expensive machines in each line have oversized bowls, and large feed tubes that accommodate whole tomatoes or fruits. They both do an excellent job of shredding and blending, though neither produces a smooth puree. They slice and chop about average, although the smaller Cuisinart chops very poorly.

The Robot-Coupe that I tested tended to "walk" when doing heavy tasks. When carrots became lodged between the slicing blade and the cover, the entire machine jumped around, and I could not operate the "off" switch. I had to hold the machine down with one hand and pull the plug with the other. This could be dangerous. This problem also occurred when I tested a bread recipe that called for 2 cups less flour than this machine is supposed to handle.

The large-size Cuisinart and Robot-Coupe have large, heavy housings that make them clumsy to handle. When the bowls are in place, they stand quite high; neither machine will fit underneath my low cupboards.

Middle of the line processors, which sell in the $80–120 range, include the Sunbeam 1451, the Hamilton Beach 702, and the General Electric FP6.

The quality of the housing and chopping and slicing/shredding blades were not as good in these machines as in the higher-priced models; however, they are adequate for the tasks they must perform—with the General Electric FP6 being of especially good quality for the price.

The General Electric FP6 has the greatest number of advantages for preserving food at home. It has a side-discharge slicing and shredding chute that allows the cook to process large quantities of vegetables at one time. Accessories include a thin-slice/fine-shred, a thick slice, and a french fry blade, as well as an oversize funnel to attach to the feed tube. The greatest disadvantage I found was that the small tool needed to reverse the disks was easily misplaced. The overall quality of the processed food was average or above. I rated this processor the best buy for the money.

The quality of the foods processed in the direct-drive Sunbeam was average or above average, and the blades are easy to handle and store. It has a good-size bowl, and the food pusher doubles as a measuring cup. The Hamilton Beach that I tested is belt-driven and did not handle heavy tasks well. It overheated, and the overload switch did not function until the belt had been chewed up. Overall quality of the processed food was average or below. I would not recommend this product for the heavy job of processing food at harvest time.

The least expensive food processor is the La Machine with the Vegetable Chef Cutting/Chopping Unit by Moulinex. This food processor has a chute discharge feature; however, the feed tube is very small. It has many small parts, which are easily lost, and the overall quality of the finished food was poor. I do not recommend this machine.

Blenders, Strainers, Food Mills, and Other Useful (and Useless) Tools

Before the electric food processor, there were many electric and manual appliances that made food processing easy. I still use my hand-cranked Squeezo strainer and wouldn't be without it. Here are a few appliances you may want to use during the harvest season.

Blenders

Blenders have become almost a standard appliance in most kitchens. They can be used to mince, chop, grate, puree, or just blend. They are not as efficient as food processors for coarsely chopping or grating vegetables, since they tend to chop too finely; however, they do a better job of pureeing than the food processor. Blenders are easier to clean than food processors. Prices range from $20–40.

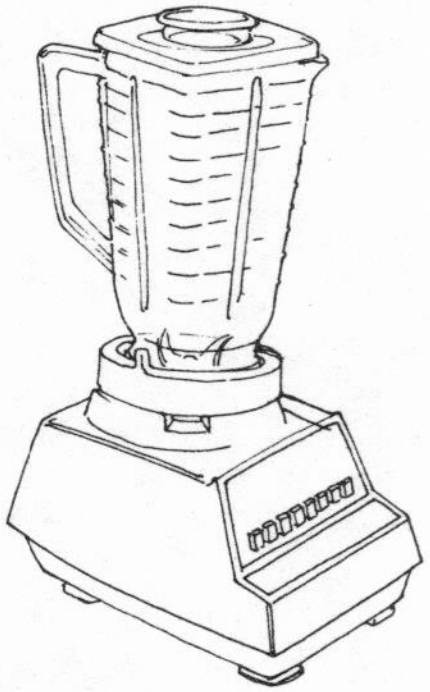

blender

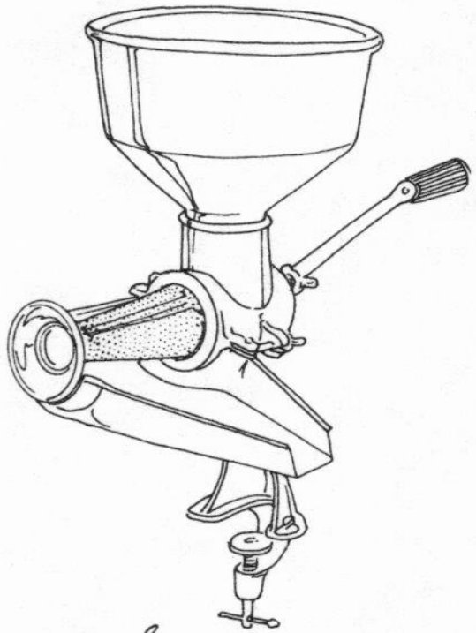

Squeezo strainer

Strainers and Food Mills

If you make your own tomato juice, sauce, or puree; make soup purees; freeze winter squash or pumpkins; or make your own applesauce, you will not want to be without a good strainer or food mill during the harvest season.

There are 3 basic types of strainers and mills; but the best if you preserve large quantities of food is the Squeezo strainer, or its almost identical twin, the Victorio strainer. They sell for $35–60.

These strainers use hand-cranked augers to crush the food and force it through perforated, cone-shaped strainers or screens. Large hoppers direct the purees into bowls, while augers move the skins, seeds, and cores into waste bowls at the other end. To process apples in these strainers, just quarter and cook them until tender. Tomatoes can be quartered and pureed uncooked. I can strain 7 quarts of tomato puree in just 10 minutes in my Squeezo, and the finished product is excellent.

These strainers are meant for large jobs, and they take a little extra time for clean-up. All parts can be washed in the dishwasher. The strainer screens, which are the hardest part to clean, can be cleaned quickly and easily by forcing a large canning jar scrub brush down inside the straining screen and giving it a few twists. Before removing the brush, scrub the outside of the screen with a small stiff-bristled brush. Both models have a straining screen for berries and a coarse screen for pumpkin.

The Foley Food Mill, shaped like a 2-quart saucepan with holes in the bottom, has a handle with a horizontal cranking arm, which operates a semicircular disk that forces the food through the holes. Underneath, a spring-loaded wire scrapes the bottom clean. Releasing a nut on the bottom allows it to come apart easily for cleaning. The Foley Food Mill doesn't work efficiently on items with tough skins. For instance, tomato skins and seeds tend to get clogged in the bottom; however, by reversing the action of the disk, it can be unclogged quite easily, and you can either remove the skins or continue straining. The 3½-quart model sells for under $20.

There are chinois strainers, cone-shaped strainers with pestles, that come in aluminum or stainless steel. A large wooden pestle, made to fit snugly in the bottom of the cone, presses food against the sides to squeeze it through. Tomato seeds are frequently forced through the holes, giving the tomato puree a speckled look. These strainers cost under $20.

Slicers

There are a variety of manually operated vegetable and cabbage slicers on the market costing from $3–35. While they can't possibly be as efficient as a food processor, they still produce a nice product with less effort than slicing with a knife.

Bean Slicers

I have discovered that my General Electric FP6 food processor with the thin-slice blade does such a good job of julienne slicing beans that I would never use anything else. I can do a bushel of beans julienne-style in 35 minutes. However, not all food processors do as good a job of julienne slicing. If I had to buy a bean slicer, I would purchase the little hand-held Krisk that sells for only $3. It is slow, but it does the best job.

The Schulte is a sturdy little $15 cast-iron hand-cranked model that clamps to the table or shelf. It does 2–3 beans at once, but they tend to tear or look ragged. The Lady Joe Combination Pea Sheller/Bean Frencher is a plastic boxlike device similar to a pea sheller. It has steel cutting blades and is driven by power from an electric mixer or variable-speed drill. Large beans tend to get stuck; however, if directions are followed carefully, it does work. It sells for under $15.

Pea Shellers

I tested 3 varieties of pea shellers: an all-metal hand-cranked model for small batches, a plastic boxlike device with stainless steel blades that uses a mixer or variable-speed drill for power, and a large aluminum sheller with a 1/10 horsepower motor.

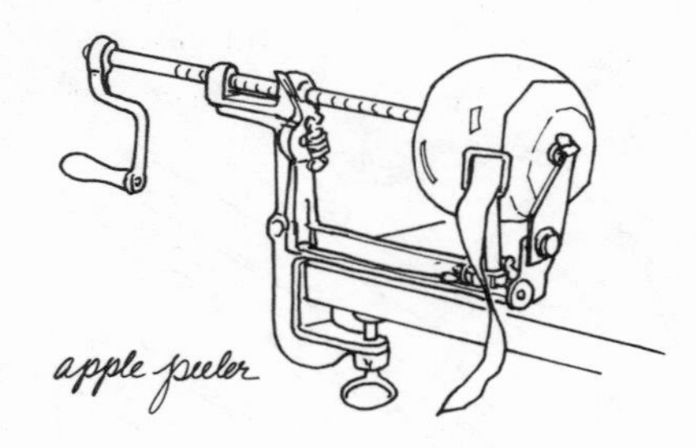

First I tested the shellers with unblanched pea pods. According to the directions for each model, only freshly picked peas have to be blanched. Peas that have been picked for more than 24 hours (and have lost all their flavor and sweetness) do not need blanching. Each unit worked on the same principle—you feed the pod between the rollers; the peas are squeezed out into a bowl or tray, and only the pod passes through. Well, the pods became jammed halfway through, some of the peas were mashed, the beater adapters on my electric mixer were stripped, and I got totally disgusted. Blanched pea pods worked well with all 3 machines.

Using blanched pea pods, the hand-cranked model produced 3½ cups in 15 minutes and was easy to clean. It sells for under $15. The power-driven Magic Finger Sheller turned out 6 cups in 15 minutes and was easy to clean. I found that it worked better with an electric drill than with the mixer. It sells for under $15. The heavy-duty power sheller I tested did a whopping 22 cups in 15 minutes and sells for $160.

Pea shellers are just a waste of money, in my opinion. The expensive model is the only one that does a good job, and even that one tends to mash the peas. Also, you have to eat an awful lot of peas to recover your investment. If shelling peas is a frustrating job for you (I enjoy the chance to sit for a change), you would do well to consider raising the Sugar Snap variety that can be eaten or frozen, pod and all.

Apple Peelers

There are several apple peelers on the market. I have tried the Nor-Pro and the Reading 78. The Nor-Pro peeled, sliced, and cored; and although the blade requires a little guidance, I found that it did a good job. The Reading, which doesn't slice or core, did not peel well, and the apples kept falling off. Other apple peelers that are rated good are the Johnny Apple Peeler and the White Mountain Apple Peeler; both slice and core. Prices are all about $20.

I consider my apple peeler a nice extra, not a necessary piece of equipment. Unless you put up bushels of sliced apples, you will not find much use for a peeler. Apples do not have to be peeled to make applesauce with a strainer.

Freezing Aids

Boilable plastic bags and automatic sealers are relatively new to the market. They have been reviewed by the FDA and been found completely safe for processing and cooking foods in. In my opinion, the boilable bags are one of the best new ideas in a long time. They can be used for freezing fresh produce, leftovers, or freshly made casseroles, soups, and stews in single-size portions to be reheated later, right in the boilable bag.

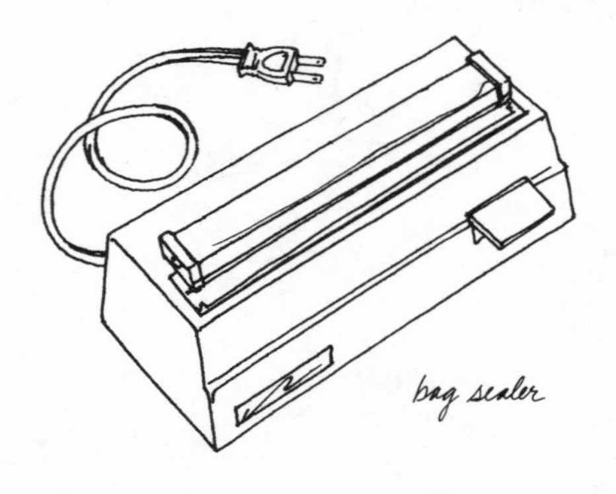
bag sealer

If your only experience with boilable bags has been with commercially frozen foods that you reheated in the bag, and you have not been pleased with the results, give these bags a second chance. As you will see in chapter 4, you can save a lot of time when freezing vegetables by packing the prepared vegetables in the bags, then blanching, cooling, and freezing—right in the bag. When it comes time to cook the vegetables, you can reheat them in the bag, but some are best steamed or stir-fried.

The boilable bags are made of heavy-duty, food-safe plastic, and most sealers work by placing the top of the bag on a small, heated sealing strip inside the appliance. Some sealers are designed to use continuous rolls so that you can custom-cut the size bag you need. There are other models that will seal different foods in separate sections of the same bag, allowing you to put together an entire meal at a time.

Some manufacturers offer a vacuum seal feature on their sealing units. These sealers are more expensive ($35–70). Although they do a good job of sealing, it is a waste of money to purchase a vacuum sealer for freezing vegetables since even the best machines cannot remove all the air.

My favorite sealer is the Oster Automatic Bag Sealer with touch control. It allows you to hold the bag with both hands; and just a touch from a finger lowers the sealing bar. Prices for sealers range from a low of approximately $10 for the Dazey Seal-A-Meal SAM 1 (always on sale) to a high of $70 for the Krupp vacuum sealer.

You do not have to invest in a bag sealer. You can fill a boilable bag, place the bag on a heavy bath towel over the edge of a counter, cover the bag with a damp cloth, and seal with an electric iron set on low. The electric iron isn't as convenient as the automatic bag sealer, but trying the process with an iron will let you know whether you want to invest in the equipment.

Blanchers and Canners

The pots in which you blanch vegetables, cook down relishes and jams, and process canned foods are necessary pieces of equipment. But before you buy anything new, be sure that these pots will be needed.

Blanchers

The traditional way of preparing vegetables for the freezer is to blanch the vegetables, cool them, drain, then bag. It is not the fastest way—using the boilable bag method is faster, as we will see in chapter 3.

There is one blancher on the market that is so handy I find many uses for it, even though I blanch most of my vegetables in the boilable bags. The 5-in-1 Pot Steam Blancher is a versatile little kettle made up of a 7-quart deep pan, a deep steam basket, a smaller colander type of basket, a flat, perforated container that holds about 2 cups, and a cover. It can be used for cooking and steam blanching (3 layers at a time), and the colander can be used for straining. It's the handiest pot in my kitchen and sells for about $20.

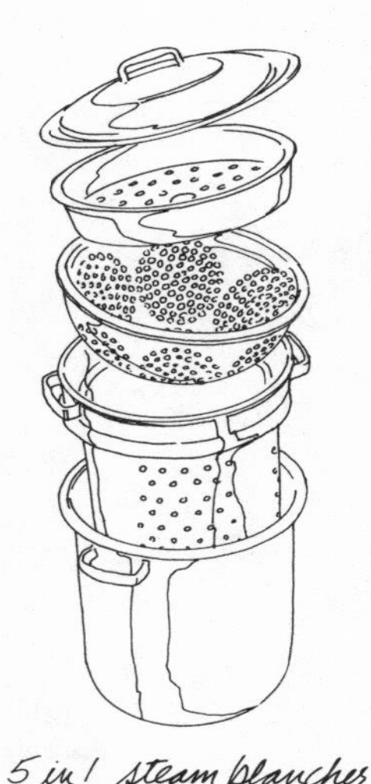

5 in 1 steam blancher

A large pan with a cover that can fit over 2 burners, such as a large roaster, is a necessary piece of equipment for the boilable bag method.

Steam Canners

Not to be confused with pressure steam canners, steam canners consist of a shallow well bottom, a perforated aluminum rack, and a dome top. About 2 quarts of water are heated to boiling in the well, up to 7 canning quart jars are placed on the rack, and steam is allowed to fill the dome. It takes about 5 minutes to fill the dome with steam hot enough to sterilize. The steam heats the canning jars and processes the food. Only acid foods can be canned in a steam canner. Steam canners make food processing much faster than boiling water bath canners; since only 2 quarts of water must be heated, overall time is much shorter.

Boiling Water Bath Canners

Also called "cold pack canners," boiling water bath canners can be used to process acid foods only. They come in 2 sizes: the smaller one can process up to 7 quarts at a time, and the larger size can process 9 quarts.

They can be made of stainless steel, aluminum, or porcelain-on-steel—the latter being the least expensive, selling for $15–25. Any large kettle can be used in place of this type of canner as long as it is as deep as a quart canning jar, plus 4 inches. It should have a cover in order to keep up a good rolling boil throughout the canning process.

Pressure Canners

Low-acid foods must be processed in a pressure canner. There are 2 types: those with a weighted gauge to regulate pressure, and those with a dial gauge. They also come in several sizes. I recommend the 22-quart size since it will hold 20 pint jars and can double as a boiling water bath canner. Pressure canners are made of heavy aluminum and prices range from $35–160—but they are an absolute must for canning vegetables and meat.

pressure canner

steam canner

Final Thoughts

If you were to ask me what equipment I considered indispensable in my harvest kitchen I would answer without hesitation: my freezer, dishwasher, food processor, Squeezo strainer, and blender. I purchased each of these appliances as soon as they became available for home use, and I can assure you that over the years, they have paid for themselves many times over.

A dishwasher is one of the best time-saving appliances that you can invest in. It will wash several batches of canning or freezing jars at once and keep them hot until needed. Also dishwashers are a help with the everyday clean-up chores that need to be done during all seasons.

For canning acid foods, pickles, and preserves, I prefer the steam canner. If you can't afford a freezer, you do need a pressure canner to preserve low-acid foods; but a freezer allows you to save time and produce a more palatable end product.

There are many more gadgets and appliances on the market. Most are worthless, but occasionally, something good comes along. If you find an item that looks like it might work for you, and you can afford it, give it a try. After all, it's the fun of experimenting that takes our daily lives out of the ordinary.

Tips and Hints

• If you cannot afford some of the larger pieces of food processing equipment, you might consider getting together with 2 or 3 friends and share the costs. This will mean that you must either stagger harvest dates or plan to process foods together in sort of a mini-cooperative.

• As you buy new kitchen appliances, put the directions for assembly and use, parts order slips, and warranties in a drawer; or make up a loose-leaf binder with them, and put it with your cookbooks. This will save time when using these appliances for the first few times, to say nothing of the aggravation that comes from looking for these papers when they are scattered all over, or when you need to have the appliance repaired.

CHAPTER 2

Tips, Hints, and Other Shortcuts

The *best* way to preserve fruits and vegetables is to pick them at the peak of their maturity and bring them from garden to canner or freezer immediately. However, for many of us this is *almost* impossible. I emphasize "almost" because with proper advance planning, you can find the time for food preservation. "But," you ask, "all the planning in the world won't prevent the green beans from ripening in the middle of the week when I have the least amount of time. Then what do I do?"

10 Harvesting Tips

If you stagger the planting dates of vegetables, they won't all be ready at the same time. This will make it easy to harvest and process small batches of vegetables frequently, in short blocks of time.

Vegetables picked late in the afternoon or early in the evening (with a few exceptions) will keep better. They have been manufacturing natural sugars and nutrients all day; once the sun goes down, they will use up part of these sugars and nutrients and are at their low ebb early in the morning.

Berries keep better if they are picked early in the day.

Pick only the best vegetables for storing. Use bruised or less perfect produce for daily meals. When picking produce, handle with care to prevent bruising.

When picking berries, collect the fruit in small containers to avoid packed and crushed fruit. As soon as each small container is filled, place it in the shade.

Small amounts of fruits and vegetables can be stored in a refrigerator or ice chest. Place berries in colanders when refrigerating.

Large plastic or metal garbage cans, with holes punched in the bottom for drainage, make good containers for extra produce when filled with ice. Place the fresh fruit or vegetables in plastic bags, then place another layer of ice on the bags, and cover with newspapers to keep in the cold. Set the can in a shady, protected spot. This will keep produce fresh for 2–3 days. Fresh ice should be added daily.

Fruits, vegetables, and large ice packs or plastic jugs of ice wrapped in plastic bags to prevent leaking (condensation forms on the outside of jugs of ice) can be stored temporarily in your dishwasher, clothes dryer, or electric oven. (Do not use gas ovens with pilot lights.) Place a heavy towel in the bottom of the appliance; add the ice, then the fruits or vegetables. The insulation that prevents heat loss from these appliances will also keep in the cold. If you must store the produce for longer than 24 hours, change the ice. Return the melted ice packs or jugs to the freezer and refreeze for another time.

Do not wash fruits and vegetables to be stored before processing. Wash when you are ready to freeze or can.

17 Preserving Tips

You will find most of my tips for saving time in chapters 3 and 4 where I talk about specific ways of preserving food. Here are just a few reminders.

Store harvested vegetables in plastic bags and cool immediately. Use several smaller bags rather than a single large one to help prevent bruising.

If you find yourself faced with mounds of vegetables and not enough time to process the lot of them, try tray-freezing them. Wash the vegetables, and dice or chop. (The smaller the pieces are, the faster they will freeze.) Spread the vegetables out on a few cookie sheets, place in the coolest part of your freezer, and allow to freeze. Once frozen, pour the vegetables into plastic bags. Chopped fresh vegetables can be frozen without blanching to be used within 2 months, if they are to be used in relishes or sauces. There is a little loss in the crisp texture of the relish, but few people will notice it. Make sure when defrosting that all the juices are collected in a bowl and used in the recipe as part of the liquid.

The fastest way to preserve berries is to wash them in cold water, *lift* the berries from the water, place in a colander, and drain well. Tray freeze without sugar. When frozen solid, pack loosely in freezer bags. These berries can be used as you would use fresh berries—for jams, jellies, pies, puddings, and sauces.

The fastest and easiest method of making applesauce is with the Squeezo strainer. Apples need not be peeled or cored, just cooked until soft and pureed in the strainer. Applesauce can be frozen or canned.

Use only firm, unblemished fruit to freeze for pies, fruit compotes, and shortcakes. Soft, overripe fruit will turn to mush when defrosted.

Overripe, unblemished fruits can be frozen sliced or whole (if small) to make quick purees when defrosted. Pop the defrosted fruits into a blender or food processor to make a smooth puree in an instant. Strain only if a very smooth puree is desired.

Another way to deal with mounds of fruits that have ripened all at once is to freeze or can them as juices to be made into jellies when you have the time.

Even if you don't have a proper root cellar, you can store pumpkins and winter squash until late in the fall or winter when you have more time for processing. Keep the squash and pumpkins on a closet floor after curing, and they will keep well for 2–3 months in these dark, cooler places.

Apples, peaches, and pears should be sliced into a bowl of water containing lemon juice or ascorbic acid to prevent darkening. When all the fruit is prepared, drain well, and tray freeze or can in syrup.

Prepare syrup for syrup-packed frozen fruits and berries ahead of time when you have a few extra minutes; it will save time when preserving the fruits.

I like to use an electric coffee maker to reheat and keep syrup hot for preserving fruit. The handle and pouring spout make it easy to fill jars. Older metal percolators should be cleaned by adding 2 tablespoons of vinegar to a pot of water and running it through a full cycle. *Caution*: Owners of automatic drip coffee makers should place syrup directly *into* coffee pots. *Do not put syrup in water well to reheat.*

To obtain a very clear jelly, strain fruit in muslin bags overnight. Do not press the fruit in the bag. If juice remains in the fruit, press the fruit in another bowl to extract the extra juice. Use this second extraction for another batch of less clear jelly.

To eliminate the problem of sugar crystals in freezer jams and jellies, measure the prepared fruit into a food

Do not hull strawberries before washing or they will absorb large quantities of water.

processor, add sugar, and blend for 30 seconds. Let the fruit stand in the food processor for the length of time specified in the recipe, add the pectin mixture, and blend again for 30 seconds.

To remove seeds from berries and grapes for jams and conserves, puree in the Squeezo strainer with the berry strainer.

The best way to defrost tray-frozen berries is to place them in a bowl and cover with whatever sweetener you will be using in your recipe; do not stir. Cover the bowl tightly. This prevents exposure to air, and the berries retain better color, texture, and flavor.

To make pies or muffins using tray-frozen fruits or berries, do not defrost before adding to the pie crust or batter. Add a few extra minutes to the baking time for muffins and 15–20 minutes for pies. This may necessitate covering the edges of the piecrust with foil for the first part of the baking time to prevent overbaked edges. Increase the amount of thickening to berry pies by half again as much. For example, if your recipe calls for 2 tablespoons of quick-cooking tapioca, add 3 tablespoons.

9 Tips for Setting Up an Efficient Work Flow

Arrange your kitchen for greater efficiency prior to the harvest season. Having a place for everything and knowing where everything is will save time, work, and frustrations.

Pack away seasonal dishes and entertainment items that take up a lot of shelf or cupboard space to free kitchen space for better harvest time use.

Plastic laundry baskets make good storage containers for out-of-season items because they can be labeled and stacked in a small area.

If possible, set up a dining area for your family in a room other than your kitchen to allow you to keep your work area set up at all times during the heaviest part of the harvest season.

If possible, bring a picnic table into your kitchen for a large work area. The tops of most of these tables are just the right thickness for clamping on food processing equipment. I have found that if you are a short person, it is less tiring to use equipment with a crank handle on these tables because they are slightly lower than the average counter.

Hook a shoe bag with deep pockets to the inside of a door off your kitchen or pantry to hold small kitchen tools.

Make yourself a cutting board to fit over your sink. Use 1-inch pine cut slightly larger than the well of your sink. In one corner, cut out a 4-inch square hole with a jigsaw. (Outlining the area to be cut with masking tape will prevent the wood from splintering while it is being cut.) Place a plastic dishpan or bowl under the hole. As you peel and chop, the waste goes directly into the waste container to save time and messy clean-ups.

To fill jam and jelly jars quickly and easily, set the jars to be filled on a lazy Susan next to your preserve kettle. Rotate the lazy Susan as you fill the jars.

Make sure you have plenty of fresh herbs and spices. Take the time to mark the date on new containers as you open them. Once opened, herbs and ground spices should be used within 6 months, and whole spices within 1 year. After that they should be discarded because they lose flavor.

Use the inside doors on the kitchen cupboards to mount spice racks. They won't collect dust and grease and will keep spices and herbs visible and quickly accessible. Line up spices and herbs in alphabetical order.

Closets are usually dark and cooler than the rest of the house, so they make good food storage spaces. Make use of the backs of closet doors for food storage. Racks are now available that will hold as many as 50–55 quart jars and cost as little as $40. Steel canning shelves that will fit in large closets will hold up to 220 quart jars. These cost as little as $35.

21 Ways to Make Life Easier During the Harvest

If gardening and food preservation are to be an important part of your life, take steps to fit them smoothly into your schedule. Many things can be done year-round to make the harvest season less frantic and tiring. These shortcuts can be incorporated into your regular work schedule, and you won't even miss the few extra minutes that most of these tasks take.

Start by obtaining a calendar with large empty squares. Plan your year-round activities dealing with food preservation, and mark the times for these activities on this calendar. Variations or cancellations will occur, but important projects won't be forgotten and will be done on time.

Fall Through Late Winter

Shop for food preservation equipment in the fall to save time and money. Food preservation equipment usually goes on sale right after the harvest season.

Take the time to teach each member of the family how to do simple jobs: set the table, clear away after meals, run the dishwasher, do the laundry, make beds, dust, vacuum, and so on. If your husband has never shopped before, take him shopping with you a few times so he can see the layout of the store and understand unit pricing. When harvest season arrives, you can just hand him a list and he can do the shopping for you on his way home from work.

Make soup stocks to store in the freezer for quick, fresh vegetable soups in the summer. This is especially good if you heat with wood or coal stoves; you can use the energy for cooking that you are already using to heat your home.

Have the gauge on your pressure canner tested at your local Cooperative Extension Service office.

Plan the types of foods you want to preserve before you plan your garden. For instance, if you plan to can

Clear at least 1 large shelf in a clean, enclosed area (kitchen cupboard, nearby closet, or pantry) to store canning jars that can be washed ahead of time.

or freeze large quantities of tomato juice, any good tomato will do. On the other hand, if catsup, spaghetti sauce, or chili sauce are desired, Italian plum tomatoes are the type you should plant. Try a few plants of Burpee's Longkeeper. These tomatoes can be harvested and left on a shelf to ripen well into the winter without spoiling.

Plan your vegetable garden, setting down approximate planting dates for various vegetables to stagger the harvest. This will prevent the problems that arise when everything is ready at the same time.

Go through your recipe files and cookbooks to find your favorite summertime recipes, especially those using vegetables that you will be preserving. Add new ones that you would like to try. Aim for easy, quick-to-prepare foods. Put these recipes in a special file.

Collect recipes for sauces, jams, jellies, preserves, pickles, and relishes (see recipe section for more). Put these recipes together in your special summertime file or create a new one for them.

Once you have decided which recipes you would like to try, calculate how much sugar, canning salt, spices, vinegar, sealing wax, and freezer bags you will need, and buy a little each time you shop. This will save extra shopping trips at your busiest time and spread the cost of the extras over a longer period of time.

Early Spring Through Early Summer

As winter slips into early spring, prepare extra meals to tuck away in the freezer for harvest day dinners without effort.

Make little muslin herb and spice bags, food straining bags, heavy potholders, and cover-all aprons. These aprons save on laundry and can be made easily from old shirts. Just remove the collars and sleeves and finish off the rough edges with binding tape. For the little ones, make large bibs out of old flannel-backed plastic tablecloths and a little bias tape.

Old stockings and panty hose make handy food strainers. Cut the crotch and legs out of clean, old panty hose, tie or sew one end, and use them for extra straining bags.

Check your canning jars for nicks and cracks, discarding those unsafe for food storage.

To save time in late summer, locate farmers' markets or truck gardens in your area in the late spring or early summer. Vegetables and fruits from these sources can supplement your own garden produce.

As your root cellar empties in the spring, clean all shelves well to avoid insect and bacteria problems during the warmer months.

Defrost and clean your freezer late in the spring. (Use a blower-type hair dryer to hasten the defrosting.) Remove all frozen foods from plastic containers and refreeze the food in marked plastic bags. The foods will come out of the plastic containers easily if you run a little hot water over the bottom and sides. Wash containers in warm sudsy water containing baking soda to get rid of unwanted odors. Throw away all foods that are past their prime, or use them as soon as possible in soups, stews, or casseroles. Those that must be discarded should go out to the pigs or to the compost pile.

Spray the walls and bottom of the cleaned freezer with a vegetable oil spray to make future clean-ups easier.

Scrub out any extra garbage cans you have, or buy a new one. In the height of the harvest season, you can fill these cans with ice and produce to keep your vegetables fresh and chilled.

To save on the cost of running your freezer, fill it with ice packs (see chapter 3, page 22) or milk cartons of water as it empties. A full freezer costs less to run, and the ice will come in handy to cool blanched vegetables.

Get together with a group of friends and encourage local stores and extension services to give food preservation classes in the late spring prior to the harvest season, rather than during harvest season when you are too busy to attend.

CHAPTER 3

The Basic Techniques of Root Cellaring, Freezing, and Canning

The easiest, fastest, oldest method of food preservation is root cellaring, or cold storage. The problems with this method are that many vegetables do not adapt well to cold storage, and many homes do not have root cellars.

The second fastest and easiest method of food preservation is freezing. But, not all vegetables freeze well, and many people don't own freezers.

Canning is probably the most time-consuming method of food preservation. But sometimes it gives the very best product. So instead of ignoring canning altogether, I will show you ways of saving time when canning—so that you can have the very best home-preserved foods.

The key to saving time when processing food is to be very organized—to have all your equipment ready, and to have your tasks sorted out (and hopefully shared with other members of the family). So here is a step-by-step method for each food preservation technique. Take the time to study the method carefully; then turn to the section on the individual vegetables for more details, and follow those brief steps to process the vegetables you have on hand—with the method that is best for you.

Root Cellaring

If you are one of the lucky people who has a storage area that will keep vegetables cool without freezing them, you can store apples, potatoes, cabbage, turnips, carrots, beets, onions, celery, winter squash, and pumpkins, without processing. Or you can temporarily store these vegetables until the rush of the harvest season and the winter holidays are over. With more space in your freezer, and many canning jars emptied, you can process squash, beets, carrots, apples, and pumpkins at your leisure.

Storing in a Root Cellar: 6 Easy Steps

1. Clean your root cellar or storage area once a year, just before the harvest season. Sweep out the area, scrub all your containers, and leave the area open to air for a few days.

2. Assemble your containers. Since different vegetables have different storage requirements—some like it cool and dry, some like it cool and moist—you will need to provide different conditions. Onions, pumpkins, and squash need dry conditions with plenty of air circulation. Root vegetables like it moist; store them in sturdy boxes, barrels, large plastic bags, or crocks. Packing fruits and vegetables in layers of dried leaves, straw, or crumpled newspapers will help absorb excess odors.

3. Harvest the best produce. Remove any injured or overripe produce and use immediately.

4. Season each vegetable as needed. For winter squash and pumpkins, this means leaving them to dry in the sun. Onions should be dried before storing. Read under the individual vegetables for specifics.

5. Pack the unwashed vegetables in suitable containers.

6. Check your stored vegetables from time to time. Remove any that are beginning to soften or show signs of spoiling. If possible, use immediately or freeze or can.

Freezing

Freezing maintains the natural color, fresh flavor, and high nutritive value of fresh foods. When properly frozen, vegetables are more like fresh than when processed by any other method of food preservation. Best of all, freezing is fast and easy.

I have been freezing garden surpluses for years; but this year it occurred to me that the old standard method of freezing (washing and preparing the vegetables, then blanching, cooling, drying, packing, and freezing them) might not be the fastest, easiest way to produce the best finished product. After all, you would not use the same recipe to cook every vegetable, would you? So I experimented with new ways of freezing and have had excellent results. Many vegetables can be frozen without blanching (although their shelf lives in the freezer will be shorter). Most vegetables can be processed in boilable bags for tremendous time savings. And greens can be stir-fried instead of blanched for a better product. I'll go over each method step-by-step, but first, a few words about organizing your work and getting your packaging material ready.

Packaging Frozen Foods

Proper packaging is absolutely necessary to prevent freezer burn, oxidation, and formation of large ice crystals.

For freezing most of the vegetables in this book, I strongly recommend using *boilable bags*. These are heavy-duty freezer bags that allow you to blanch, cool, freeze, then cook your vegetables right in the bag. They are sealed with an automatic sealer or your own electric flat iron.

Other freezer packaging materials you can use are freezer paper, heavy foil, wide-mouth freezer jars with straight sides (for easy removal of frozen foods), rigid plastic containers, and plastic freezer bags. Square freezer jars or rigid containers make better use of freezer space than round ones. If you don't use boilable bags, use only plastic bags that are labeled *food safe*.

It is important to remove as much air as possible from packages of food before freezing. Excess air in the packages causes oxidation, which lowers the quality of the food, and vacuum-packed bags take up less freezer space.

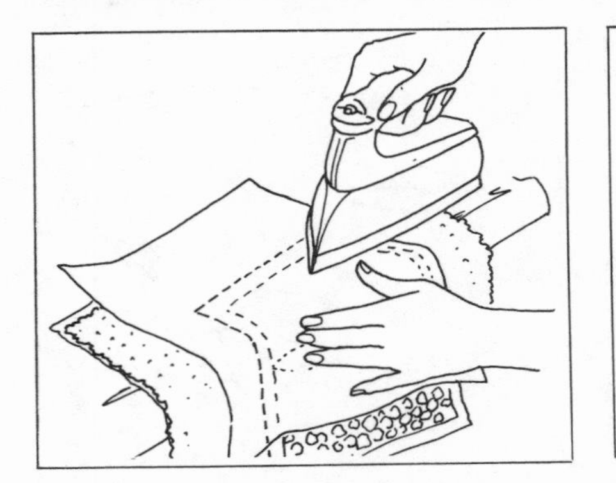

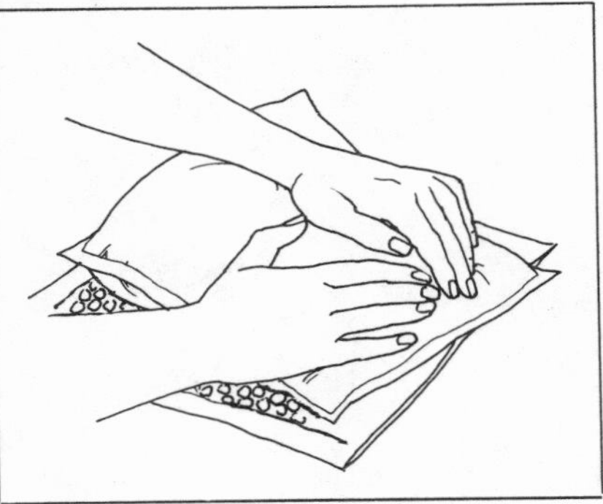

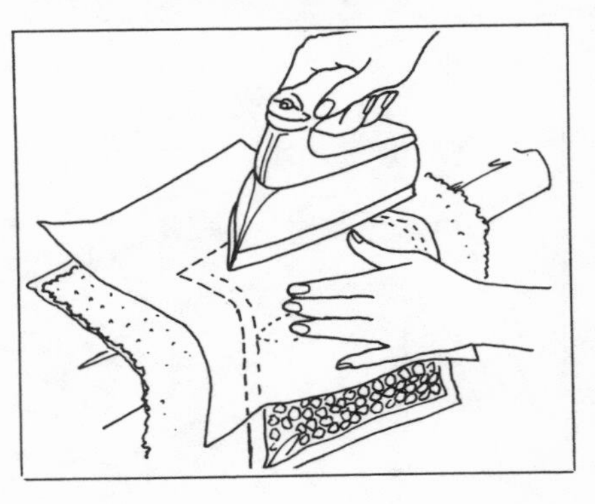

There are many accessories on the market for removing air from plastic bags, but you can remove as much air as these appliances do, without investing any money, by pressing the air out with a pillow or your hand. This method works best with boilable bags.

1. Fill the bag with vegetables and distribute the vegetables evenly in the bag.
2. With a small pillow or your hand, press down gently, but firmly, on the bag to expel all the air you can.
3. Hold the bag closed between your thumb and forefinger. Press the sealing bar down and hold a few seconds longer than recommended. Move the bag ½ inch and seal again.

If you don't have an automatic bag sealer, use an electric flat iron.

1. Place a filled bag on a heavy towel, and place a damp cotton cloth over the edge of the bag.
2. Partially seal the bag with an iron, leaving 1½ inches open.
3. Remove the air by pressing a small pillow or your hand on the bag, gently but firmly. Finish sealing with the electric iron over the damp cotton cloth.

To prevent injury when slicing vegetables with a manually operated rotary slicer, blade slicer, or slaw slicer, wear a clean cotton garden glove on the hand that is apt to come in contact with the slicing blade.

Washing Vegetables

Wash your vegetables in plenty of water. Use a medium-stiff bristled brush or a plastic or nylon net scrubber that can get into crevices where the dirt is hardest to remove. Be especially thorough with root crops since botulism bacteria may be in the soil, and only thorough washing will remove them.

Always lift vegetables out of the water rather than letting the water drain off. Drain vegetables on absorbent toweling, patting as dry as possible.

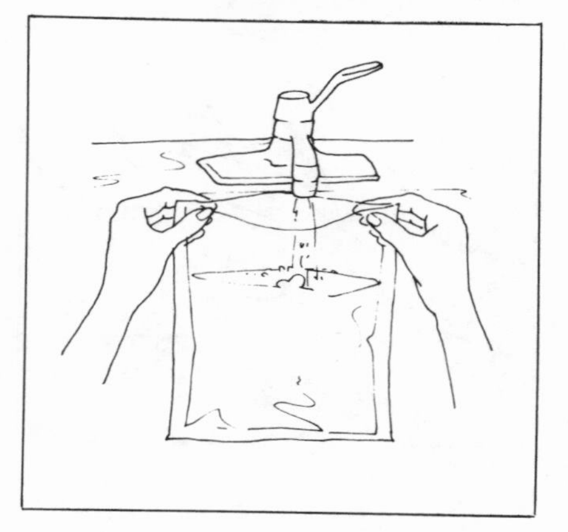

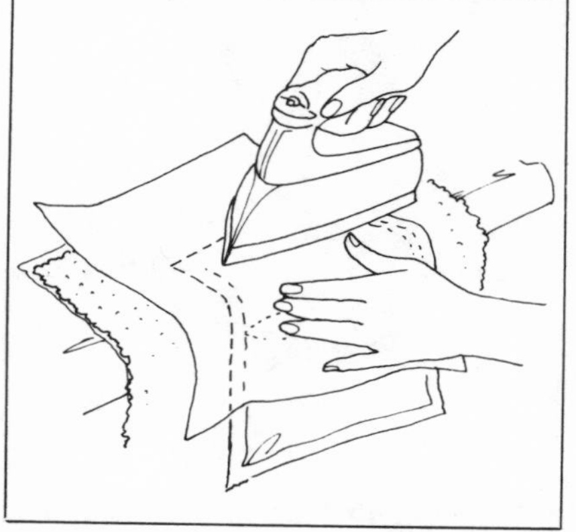

Reusable Ice Packs: A Big Time-Saver

It's worth investing in a box of quart-size freezer bags to make ice packs for freezing. These bags of ice last much longer than either ice cubes or chunks of ice. When you are through freezing for the day, remove the bags from the water, wipe them dry with a towel, and return them to the freezer to refreeze. This saves the fuss of making ice cubes and the frustration of running out of ice in the middle of a large freezing project. Here's how you can make these ice packs.

1. Fill boilable freezer bags ¾ full of cold water.
2. Hold the bag over the edge of your shelf or counter.
3. Seal with electric sealer, or place the bag on a heavy towel, cover with a damp cloth, and press with a hot iron. Freeze.

All vegetables that are blanched and then bagged to be frozen can be tray frozen, then bagged. This allows you to postpone packaging until the next day, when you may have more time. Also, it makes it possible to package vegetables loosely in large bags or containers. When you want some vegetables, they will pour freely from the bag.

Frozen unblanched vegetables are best cooked by stir-frying. To do so, melt 1 tablespoon of butter per serving in a heavy, preheated skillet. When the butter has melted, add the frozen vegetables and stir and toss the vegetables over high heat to the desired degree of tenderness. Cook until all moisture is evaporated. If more moisture is needed to cook to desired stage of doneness, add water, 1 tablespoon at a time.

Tray-frozen vegetables should be packaged as soon as possible, but if you are short on time, slip a large food-safe plastic bag over the tray and tie it closed. Package as soon as you have a chance. Do not leave trays in bags longer than 24 hours.

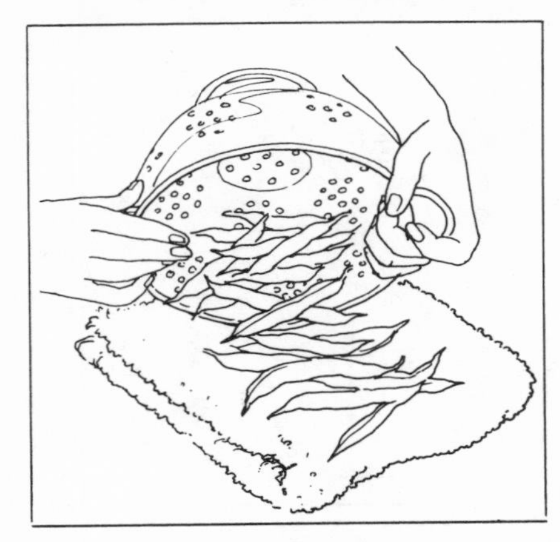

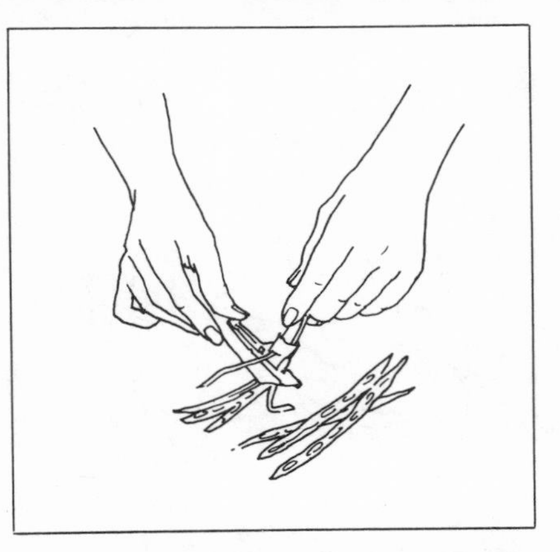

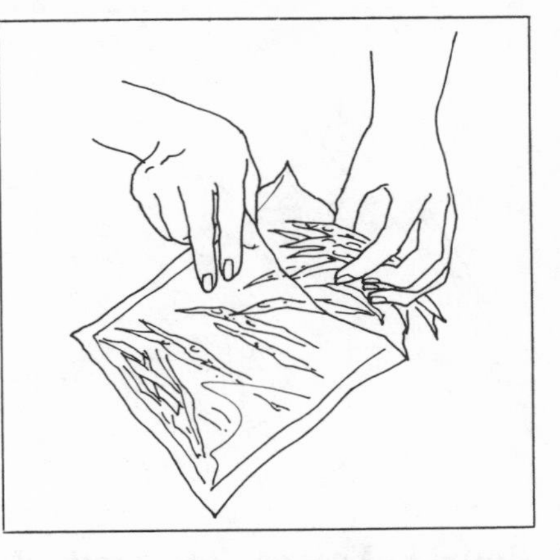

Unblanched Freezing: 5 Quick Steps

This is the fastest, easiest method of freezing. It was originally thought that this method was acceptable only for chopped onions, peppers, fresh herbs, or other vegetables that were to be stored for less than 1 month. But I have found that many vegetables can be stored for much longer and still maintain good color, flavor, and texture. Try this method with onions, peppers, herbs, celery, corn in husks, cabbage, Sugar Snap peas, summer squash, young, tender broccoli, and green and yellow beans.

1. Make sure your work area and all equipment are spotlessly clean. Assemble your equipment and set your tools where they will be most useful. You will need a scrubbing brush, 2–3 heavy bath towels, freezer bags and small pillow (optional), and a labeling pen and tape.
2. Select vegetables that are slightly *immature*. Wash the vegetables and drain on absorbent toweling.
3. Prepare the vegetables: slice, dice, chop, julienne, or leave whole.
4. Pack in freezer bags, expelling as much air as possible. Label with name of product and date.
5. Freeze in a single layer in the coldest part of the freezer.

Vegetables frozen this way should be used within 3 months. The best methods of cooking vegetables frozen in this manner are stir-frying or steaming.

The more accessible the proper tools for food processing are, the faster, easier, and more enjoyable the job will be. After using large appliances, such as food processors and Squeezo strainers, I always wash and set them back up again. I cover my Squeezo with a plastic bag to prevent dirt and bacteria from collecting in the open funnel top.

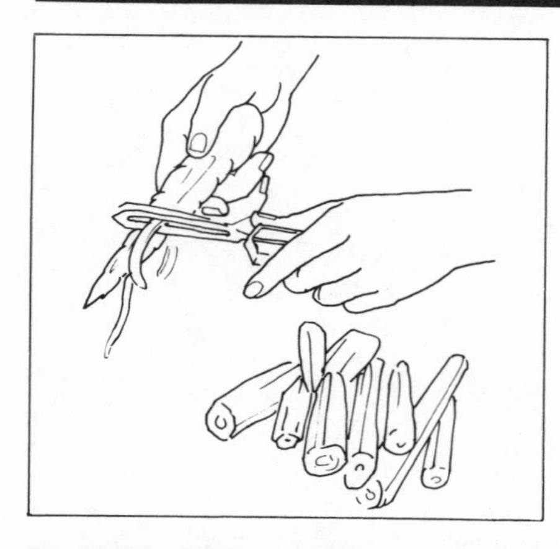

Boilable Freezer Bag Method: 11 Steps

This method of freezing often produces the best-tasting vegetables. Since the vegetables never come in contact with water, all color, flavor, texture, and most nutrients are preserved. Adding butter to the bag, when desired, coats the vegetables with a protective film that further enhances the quality and flavor of the finished product. Experiment with combinations of vegetables, such as peas and tiny onions, or peas and carrots. Sliced, diced, or julienne vegetables work best. Whole carrots and beets do not freeze well by this method. Strong-flavored vegetables, such as broccoli, cauliflower, cabbage, and turnips, should not be frozen by this method.

Time is saved with this method because bags of food can be blanched in multiples; cooling requires no special timing or handling (allowing you to continue packing); and since all vegetables are processed within the package, pans need only to be rinsed and dried, making clean-up a snap.

I tested this method against the standard freezing method with green beans. After the initial washing and trimming (time for both was the same), I timed the balance of the freezing procedure. I was able to pack, blanch, and cool ½ bushel of green beans by the boilable freezer bag method in 29 minutes, versus 1 hour and 25 minutes for the standard method. Try it yourself!

1. Make sure your work area and all equipment are spotlessly clean. Assemble your equipment and set your tools where they will be most useful. You will need a scrub brush, heavy bath towels, chopping board, knives, food processor, wide-mouth funnel or large spoon, freezer bags, small pillow (optional), bag sealer or electric flat iron, indelible marking pen and freezer tape, large roaster for blanching half-filled with water, large kettle for cooling, tongs, potholders, and a timer.

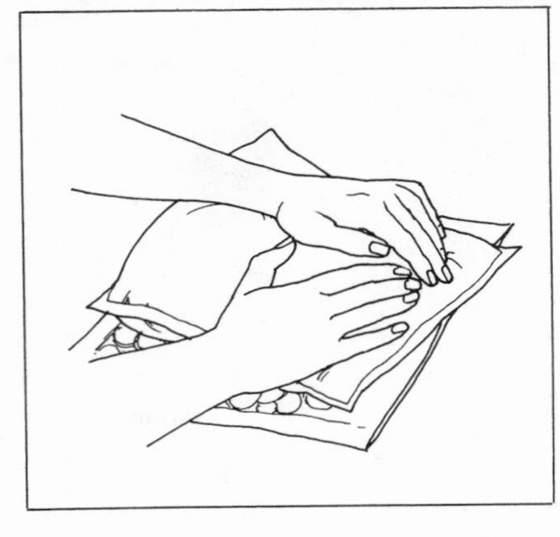

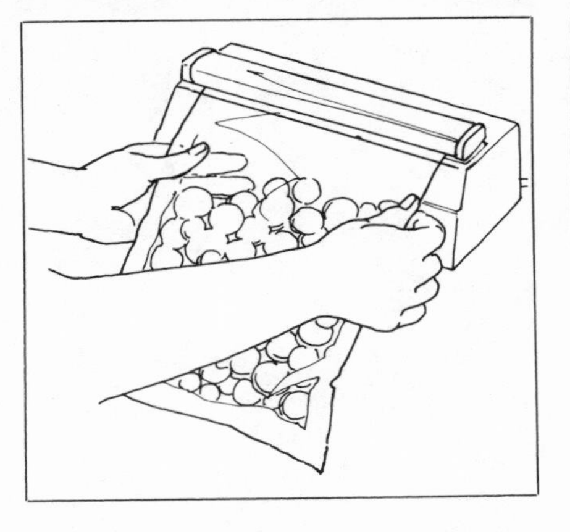

2. Select young, fresh vegetables that are just table-ready or slightly immature. Wash well, drain on absorbent toweling.

3. Begin heating water in the roaster for blanching.

4. Prepare vegetables as desired: slice, dice, chop, julienne, or leave whole (except for large dense vegetables, such as carrots and beets).

5. Plug in bag sealer or iron. Fill 4 boilable bags with vegetables in meal-size portions, making sure that when the vegetables are distributed, the package is no thicker than 1 inch. Add butter and seasonings if desired. Use as much as you would use if cooking for a meal.

6. Expel as much air as possible and seal with an automatic bag sealer. Or place the bag in a heavy bath towel, cover with a damp cloth, and seal with an electric iron. Label and date bag.

Instead of blanching your favorite vegetables, try stir-frying them in a little butter for a few minutes. Then package the vegetables in boilable bags. To prepare these vegetables for the table, reheat the bagged vegetables in boiling water. You will enjoy the stir-fry flavor and texture.

7. When 4 bags are packed, drop them into the boiling water and blanch with the pan covered. (The bags will float on top of the water. This is all right, as long as each bag has one side in contact with the boiling water.) Check the timing (see page 126) and set the timer. A rule of thumb is to blanch for double the length of time suggested for the standard blanching method. Use the slightly shorter time for tender young vegetables, and a slightly longer time for more mature ones. Start counting the time as soon as you replace the cover.

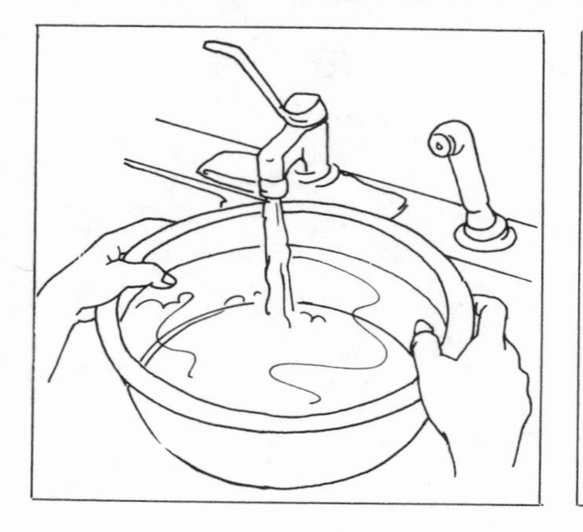

Fill the cooling kettle with cold water and ice packs.

Continue packing while the first batch is blanching.

8. When the blanching time is up, remove the packages from the blanching kettle to the ice water. Make sure the ice holds the bags down in the water, since air left in the bag tends to make the bags float. During the chilling time, occasionally knead the bags to move the cold into the center of the packages.

9. Add 4 more bags to the blancher and continue as before, until all the vegetables are blanched. Leave the processed bags in the ice until you are completely finished, unless you need the space for more vegetables.

When all the vegetables have been blanched, leave the bags in the ice water for an additional 10 minutes. While the bags are cooling, clean up your work area.

When pouring large quantities of boiling water into the sink, run the cold water. This will prevent the steam from scalding you.

10. Remove the bags from the water and dry with an absorbent towel.

11. Freeze the vegetables in a single layer in the coldest part of the freezer. Remove the ice bags from the cooling kettle, pat dry, and return to the freezer.

Foods frozen this way can be cooked in the bag or removed and steam or stir-fried for faster cooking.

To save confusion and mistakes when working with seasonings, always work from the right to left or vice-versa. As you use a seasoning, move it to the other side.

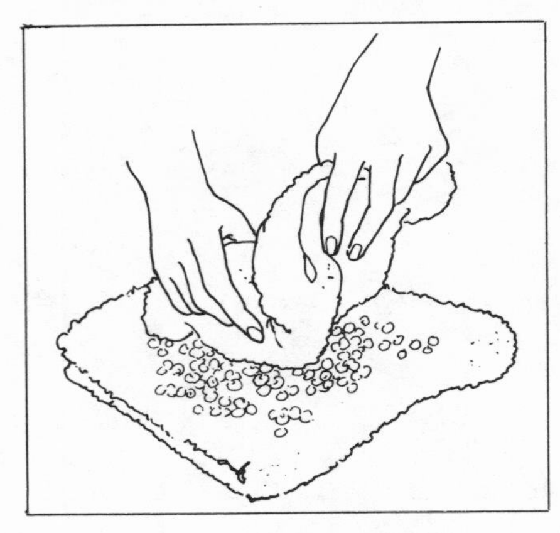

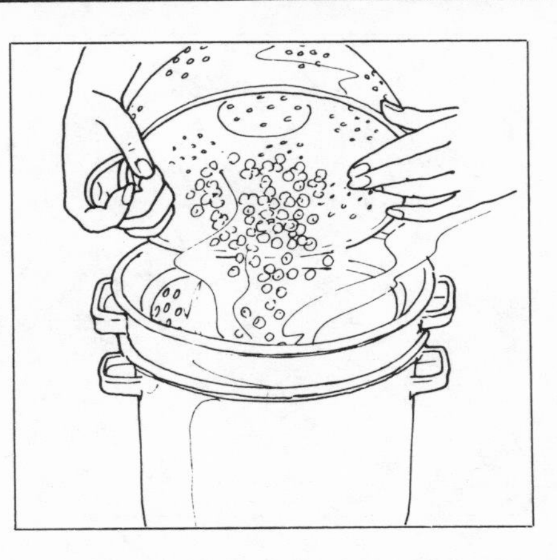

Standard Freezing Method: 10 Steps

Pick up any book on food preservation written before the days of boilable bags, or written without concern for time saving, and you will find this method. It works just fine, you get a very acceptable end product; but you will spend more time with it.

1. Make sure your work area and all equipment are spotlessly clean. Assemble your equipment and set your tools where they will be most useful. You will need a scrub brush, colander, strainers, paring and chopping knives, cutting board, food processing equipment (optional), measuring cups, tongs, heavy towels, wax-paper-lined cookie sheets, freezer containers or bags, freezer tape and indelible marking pen, blanching kettle half-filled with water, potholders, and a timer.
2. Select young, fresh vegetables that are just table-ready or slightly immature. Wash, and drain on absorbent towels.
3. Begin heating water in the blanching kettle.
4. Prepare the vegetables as desired: slice, dice, chop, julienne, puree, or leave whole.
5. Clean the sink and fill with ice water.
6. Blanch the vegetables. If you have a boiling water blancher, immerse the vegetables in boiling water, 1 pound at a time. Start counting as soon as the water returns to a boil, and blanch as long as the chart indicates on page 126. If it takes longer than 2 minutes for the water to return to a boil, blanch fewer vegetables at a time.

If you have a steam blancher, blanch 1 pound of vegetables at a time in a steam basket or blancher suspended over boiling water, for the same length of time as for boiling water. Start counting time as soon as you cover the pan. You can stack 3 blanchers to process up to 2½ pounds of vegetables at a time. Add 2 minutes to the blanching times, and begin counting time as soon as the pan is covered. It is a good idea to put the larger pieces of vegetables on the bottom and the smaller pieces at the top.

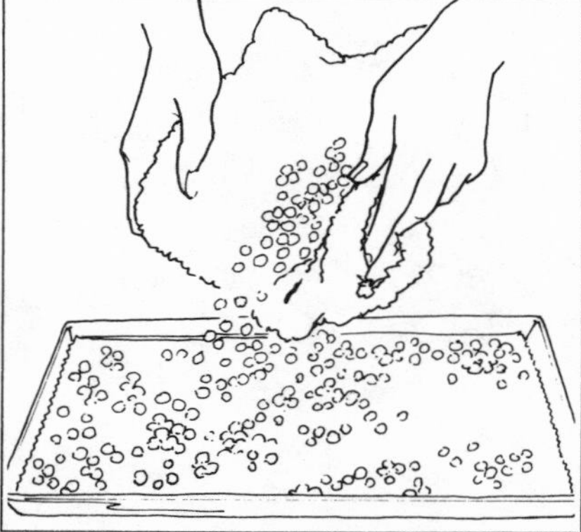

7. Cool the vegetables quickly in ice water. Cooling time is approximately the same as blanching. *Do not immerse the blanching utensil in the ice water*—this will warm the water unnecessarily, requiring more ice or a longer cooling time; and when you return the blancher to the boiling water, it will require more time and energy to return the blancher to the boiling point.

8. Drain the vegetables thoroughly, removing as much water as possible by lifting them from the ice water onto heavy bath towels and patting them dry with another towel. While one batch of vegetables is chilling, pack any previously drained batch, and blanch another.

9. Place cooled, drained vegetables on wax-paper-lined cookie sheets and freeze until solid. Then package loosely in plastic bags that have been labeled with the date and product. Be sure to remove as much air as possible. You can package these vegetables in large bags and remove just the number of servings needed at a time.

You can also package the vegetables in freezer containers or freezer bags.

10. Freeze vegetables in a single layer in the coldest part of the freezer.

Clean up when all vegetables are in the freezer. Wipe ice packs dry and return to the freezer.

Canning

Many of you will choose to preserve vegetables by canning because you don't have the freezer space. Though more time-consuming than freezing, canning is not difficult; and with proper care for cleanliness to avoid bacteria, you can produce a very fine product. In fact, whole tomatoes and tomatoes for stewing are best canned.

We will present 3 methods of canning—steam canning, the fastest method, but limited to fruits, tomatoes, and pickles; boiling water bath canning, also limited to high-acid fruits and vegetables; and pressure canning, the only acceptable method for low-acid vegetables. We will also show you 2 methods of packing jars—hot pack and cold pack.

But first, let's talk about equipment.

Canning Jars and Lids

Use only ½-pint, pint, 1½-pint, quart, and ½-gallon jars made especially for canning. Commercial jars, such as the type mayonnaise comes in, are too thin and will not withstand the heat required for processing vegetables. European-made jars cannot be used with our time charts because they are calibrated in metric sizes. Make sure your jars are free from nicks and cracks.

There are 3 types of jar closures: modern 2-piece screw bands and lids, wire bails with glass lids, and zinc caps with porcelain liners.

two-piece
screw band and lid

Modern 2-piece Screw Bands and Lids. The most popular and safest type of jar seal, these must be used with threaded jars. The metal lids have a flanged edge with a rubberlike sealing compound on their edges, which seals to the edge of the jar. The lid is held in place with a screw band during processing. After the jars have sealed, the screw band should be removed. The band can be reused to process other jars; but the lids are not reusable.

Wire Bails With Glass Lids. These old-style jars are not sold anymore, except at garage sales; however, rubber jar rings are still sold to fit them. (New modern designs of these jars are on the market, but these jars are sold mostly for storage.) Be sure to follow the directions carefully for sealing these jars, since it is more difficult to tell if they have sealed properly. Jars with loose clamps will not provide an adequate seal. Jar rings cannot be reused.

Zinc Caps With Porcelain Liners. Another old-style lid has a zinc cap that is protected from the food by a porcelain liner. If the liner is chipped, the cap should be discarded. These caps are used with rubber jar rings and must be used with a threaded jar. Follow directions carefully when canning with these caps. Rubber jar rings cannot be reused.

Preparing Jars and Lids

Wash enough jars for several batches of food all at once. A dishwasher comes in handy for this job. Place the jars upside down on clean towels in a clean area. They will be ready when you have time to can.

If your jars have not been prewashed, wash and rinse them when you are setting up. Check for chips and cracks that might have occurred while you were washing.

To heat jars and keep them hot for canning, fill the well of a steam canner with hot tap water. Place the jars upside down on the rack. Turn on the heat and bring the water in the canner to a boil. Then turn off heat but leave the jars in the canner. To sterilize jars and lids at the same time in a steam canner, place a lid with the seal up on the upturned bottom of each jar in the steam canner and steam for 15 minutes. You can also heat jars by keeping them in hot water, in a low oven, or in the dishwasher on the dry cycle.

Place jar lids and screw bands in water according to the manufacturer's directions. Place glass jar lids and rubber jar rings or zinc caps and rubber jar rings in hot water until needed.

Tips on Washing Vegetables

Wash vegetables in plenty of cold water. Use a medium-stiff bristled brush or a plastic or nylon net scrubber that can get into the crevices where the dirt is hardest to remove. A stiff-bristled brush will just skim over these areas. Be especially thorough with root crops since botulism bacteria may be in the soil, and only thorough washing will remove them from the vegetables. Always lift vegetables out of the water rather than letting water drain off, otherwise the dirt will just be redeposited on them.

Cold Pack Versus Hot Pack

Cold pack is sometimes referred to as raw pack—it means packing prepared vegetables and fruit into the canning jar raw. This is the fastest method of packing since it does not require heated jars or precooked food. Clean-up is faster because you do not have the extra pans and utensils to wash.

Any disadvantages to cold packing? There will be some shrinkage, and occasionally the fruits or vegetables will float to the top of the jar (especially tomatoes). The jars will hold slightly less when packed this way. But I don't think these drawbacks outweigh the ease of preparation and the time savings of this method.

To hot pack a jar, precook your food in simmering water, juice, or syrup for about 5 minutes; then pack the hot food loosely into hot jars. Obviously, this method is more time consuming. With tomatoes, however, the processing time for hot packing is 25 minutes less than for cold packing. You will avoid the problem of floating tomatoes, but the whole tomatoes get crushed with this method. The overall time to prepare tomatoes is still approximately the same.

For saving time, I recommend cold packing.

Sealing Jars

With the modern 2-piece screw bands and lids, you will have to buy new lids each time you process. Follow the manufacturer's directions for sealing.

To seal with a wire bail and glass lid, place a new rubber jar ring over the jar so that it rests on the shoulder of the jar; make sure you stretch the ring as little as possible. Place the glass lid on the jar and pull the longest of the 2 wire bails so that it rests in the groove of the glass lid. Leave the small bail up while the jar is processing.

To seal zinc caps with porcelain liners, place a new, wet jar ring over the jar so it rests on the shoulder of the jar, stretching only enough to make this possible. Screw the cap on firmly, then unscrew about ¼ inch to allow air to vent from the jar during processing.

When processing time is completed, remove the jars from the canner and place them several inches apart on a heavy towel. Do not tighten screw bands on jars with the 2-piece screw bands and lids. To seal jars with glass lids and bail, pull the small bails

Salt is added to canned vegetables for taste. It is not necessary for any other reason.

down. To seal jars with zinc caps, very gently retighten the lids as soon as the jar is removed from the canner.

Do not open any jar, even if liquid has boiled out. The food is safe as long as the jar seals properly.

Do not cover the jars while they are cooling. Allow the jars to cool for 24 hours; then test the seal.

Testing Seals

After the cooling period is completed, remove the screw bands on jars with the 2-piece lids to prevent them from rusting onto the jars and to help you detect broken seals.

What does a good seal look like? The lid should be depressed in the center, and if you pick the jar up by the edge of the lid, it will not come loose. With the jar tipped upside down, check for any leaks. If there are any leaks, or if the center of the jar lid pops back, refrigerate that jar to use within a few days, or reprocess.

To reprocess, clean the rim of the jar. Add liquid, if necessary, to maintain the proper head space, and put on a new lid and screw band. Then reprocess as before, for the full length of time recommended.

Storing Jars

Store jars in a cool, dark place, where they will not freeze. Warm temperatures cause discoloration and taste changes, and could cause spoilage that would not otherwise occur. Freezing does not make food unsafe, unless the jar is cracked or the cap loosened; but some foods may soften in texture.

Before using a jar of home-canned food, check for signs of spoilage. Bulging or unsealed lids, spurting liquid, mold, malodorous contents, or a constant stream of tiny bubbles when the jar is turned upside down indicate spoilage. Some bubbles are normal; but if after the contents have settled, you can see a constant flow of tiny bubbles from around the lid of the jar, then you must discard the contents. Discard the food from any jars that look suspicious. Food from these jars should be discarded where humans or animals will not eat them. The jars can be reused. Wash them well with hot sudsy water and boil them for 15 minutes before using them again.

Caution: Home canned vegetables should be heated to boiling and cooked, covered, for 15 minutes. If the food foams, looks spoiled, or smells bad when heated, do not eat it.

Make use of leisure time by chopping and dicing vegetables while watching your favorite show on television. Make a 12-inch cutting board out of 1-inch pine to fit your lap, and put the diced vegetables in a large plastic bag as you dice or cut them. Divide them into smaller bags later, or tray freeze and package in larger bags.

Acid foods, including all fruits, tomatoes, rhubarb, sauerkraut, and pickled foods, can be safely processed in a boiling water bath or steam canner. Low-acid foods include all vegetables, except tomatoes, and all other foods that do not fall into the acid group, such as meat. Low-acid foods must be processed in a pressure canner.

Steam Canning: 13 Steps

Steam canning is the fastest method of canning *acid foods*. In my kitchen, I have canned hundreds of jars of pickles, tomatoes, and fruit using a steam canner and have had excellent results.

With the steam canner, I usually save about 15–20 minutes of my time per batch. Although processing times for steam canners are the same as for boiling water bath canners, there is less water—2 quarts—to bring to a boil and keep boiling. It takes about 5 minutes to fill the steam canner dome with steam hot enough to process. The average length of time the boiling water bath takes to return to a boil after jars have been added is 25–30 minutes, depending on whether you are processing pints or quarts.

But a word of caution is in order. The USDA does not recommend using a steam canner for home processing. We have heard reports of problems with the steam canner—bad seals and broken jars. But none of these problems has cropped up in my kitchen. Through trial, error, and research, I have found a successful method of steam canning that differs a bit from the manufacturers' suggestions. Basing my method on some research findings of the University of Massachusetts at Amherst ("The Thermobacteriological Effectiveness of the Steam Canner Method of Processing Acid and High Acid Foods" by Edward A. Laperle), I wait until steam flows out of the canner in a steady stream for *5 minutes* before I start counting my processing time. This way I am certain that the temperatures inside the dome are hot enough to process the food. I think if you follow my careful method you too will have excellent results with a steam canner.

1. Make sure your work area and all equipment are spotlessly clean. If your jars have not been prewashed, wash and rinse jars and check for nicks and cracks. If you plan to hot pack, keep the jars hot in water, or in the steam canner.

Assemble your equipment and set your tools where they will be most convenient. You will need a scrub brush, colander and/or strainer, paring and chopping knives, cutting board, food processing equipment, measuring cups and spoons, nonmetallic spatula to expel bubbles (wooden chopsticks work well), wide-mouth funnel, large bowls, canning jars, steam canner, preserving kettle, tea kettle filled with water, canning lids and screw bands in hot water, large wooden and slotted spoons, soup ladle, jar lifter or tongs, timer, hot pads, mitts, or heavy potholders, and heavy bath towels.

The rack that came with my steam canner has plenty of holes through which the steam rises. But some models have just a few evenly spaced holes. If your steam canner is this type be sure never to set a jar over any hole. Otherwise you may end up with some broken jars.

2. Select perfect fresh vegetables at the peak of their maturity. Wash well, drain.
3. Prepare vegetables as desired: peel, slice, dice, chop, puree, julienne, or leave whole.
4. Fill the well of the steam canner with hot tap water. Set the steamer rack in place. Place the canner on the stove and turn the heat on high.
5. Pack the jars.

Cold Pack. Firmly pack clean unheated jars with vegetables. If desired, add ½ teaspoon salt to pints, 1 teaspoon to quarts. Fill with hot (not boiling) water or vegetable juice, leaving proper head space.

Hot pack. Place clean, prepared fruit, tomatoes, juice, or puree in a large pan; cover fruit with water or juice. Bring to a boil and simmer tomato products for 5 minutes, fruit for 3 minutes. Pack hot tomato products loosely into clean, hot jars, adding ½ teaspoon salt to pints and 1 teaspoon to quarts, if desired. Fill jars of whole tomatoes or fruit with hot liquid, leaving the head space recommended with each individual product.

6. Expel air bubbles in the jar by running a nonmetallic kitchen utensil gently between the vegetables and the jar. Add more liquid if necessary.
7. Wipe the rim of the jars with a clean, damp cloth, and seal. Place the jars on the rack in the preheated canner. Set the dome cover in place.

To remove stuck screw bands, wring out a cloth in hot water, then wrap it around the band for a minute or so to help loosen it.

8. As soon as steam rises from the dome (about 5 minutes), give the dome another 5 minutes to fill with steam that is hot enough for sterilizing. Then start counting time. Process for the same length of time as for boiling water bath processing.

The heat may be turned down during processing, but the water must remain boiling.

Two quarts of water should last about an hour of processing. Do not allow the steam canner to go dry; add boiling water if necessary.

9. If you are canning more than 1 batch of food, prepare the second batch while the first is processing. When the last batch is processing, clean your work area.

10. When the processing time is up, turn off the heat. Using heavy potholders or mitts, carefully lift the cover of the canner so that the steam is directed away from you.

11. After 1 minute, remove the jars. Place them several inches apart on a heavy bath towel, away from drafts. Complete the seals, if necessary.

12. Cool the jars for 24 hours. After the cooling period is completed, remove screw bands and check seals. Wipe sealed jars with a clean, damp cloth. Label clearly with date and product.

13. Store jars in a cool, dark place.

To prevent hard water mineral stains on jars and build-ups in canners, use ½ cup vinegar per cannerful of hard water.

Boiling Water Bath Canning: 14 Steps

Boiling water bath canning is the USDA's recommended way of canning acid foods—fruits, pickles, and tomatoes, the same foods that can be processed in a steam canner. With a boiling water bath canner, you immerse jars of food in hot water to cover and begin counting the processing time once the water begins to boil. This usually takes 25–30 minutes, depending on whether you are processing quarts or pints.

1. Make sure your work area and all equipment are spotlessly clean. If your jars have not been prewashed, wash and rinse them. Check the jars for nicks and cracks. If you plan to hot pack, keep the jars hot in water.

Assemble your equipment and set your tools where they will be most convenient. You will need a scrub brush, colander and/or strainer, paring and chopping knives, cutting board, food processing equipment, measuring cups and spoons, bubble expeller (wooden chopsticks work well), wide-mouth funnel, large bowls, canning jars, boiling water bath canner, preserving kettle, tea kettle filled with hot water, canning lids and screw bands, large wooden and slotted spoons, soup ladle, jar lifter or tongs, timer, hot pads, mitts, or heavy potholders, and heavy bath towels.

2. Place the jar lids and screw bands in water according to the manufacturer's directions. Place glass jar lids and rubber jar rings or zinc caps and rubber jar rings in hot water.

3. Select perfect fresh fruits and vegetables at the peak of their maturity. Wash the produce. Drain.

4. Fill your canner with about 4 inches of water for pints and 4½ inches for quarts. (These amounts are for a 20–21 quart canner.) Place the canner on the stove and begin heating. Heat extra water to boiling in a tea kettle.

5. Prepare the vegetables: peel, slice, dice, julienne, leave whole, or puree.

6. Pack the jars.

Cold pack. Firmly pack clean, unheated jars with fruits or vegetables. If desired, add ½ teaspoon salt to pints, 1 teaspoon to quarts. Fill the jars with hot, not boiling, water, syrup, or juice, leaving the proper head space.

Hot pack. Place prepared tomatoes, fruit, puree, or juice in a large pan; cover fruit with syrup or juice. Bring to a boil and simmer tomato products for 5 minutes, fruit for 3 minutes. Pack the hot tomato products or fruit loosely into clean, hot jars, adding ½ teaspoon salt to pints and 1 teaspoon to quarts, if desired. Fill jars of whole tomatoes or fruit with hot liquid, leaving the head space recommended with each individual vegetable.

7. Expel air bubbles in the jar by running a nonmetallic kitchen utensil gently between vegetables and jar. Add more water if necessary to maintain the proper head space.

8. Wipe the rim of the jar with a clean, damp cloth and seal.

9. When the jars are packed and the lids are in place, check the canner. If the water is boiling, add 1 quart of cold water to reduce the temperature enough to cushion the shock of adding the cooler jars to the boiling water.

Carefully lower the jars, using long-handled tongs or a jar lifter, into the hot water in the canner. Be careful not to bump jars against one another as you lower the rack; this could cause jars to crack. Water should cover jars by 2 full inches. If necessary, add more boiling water to the canner.

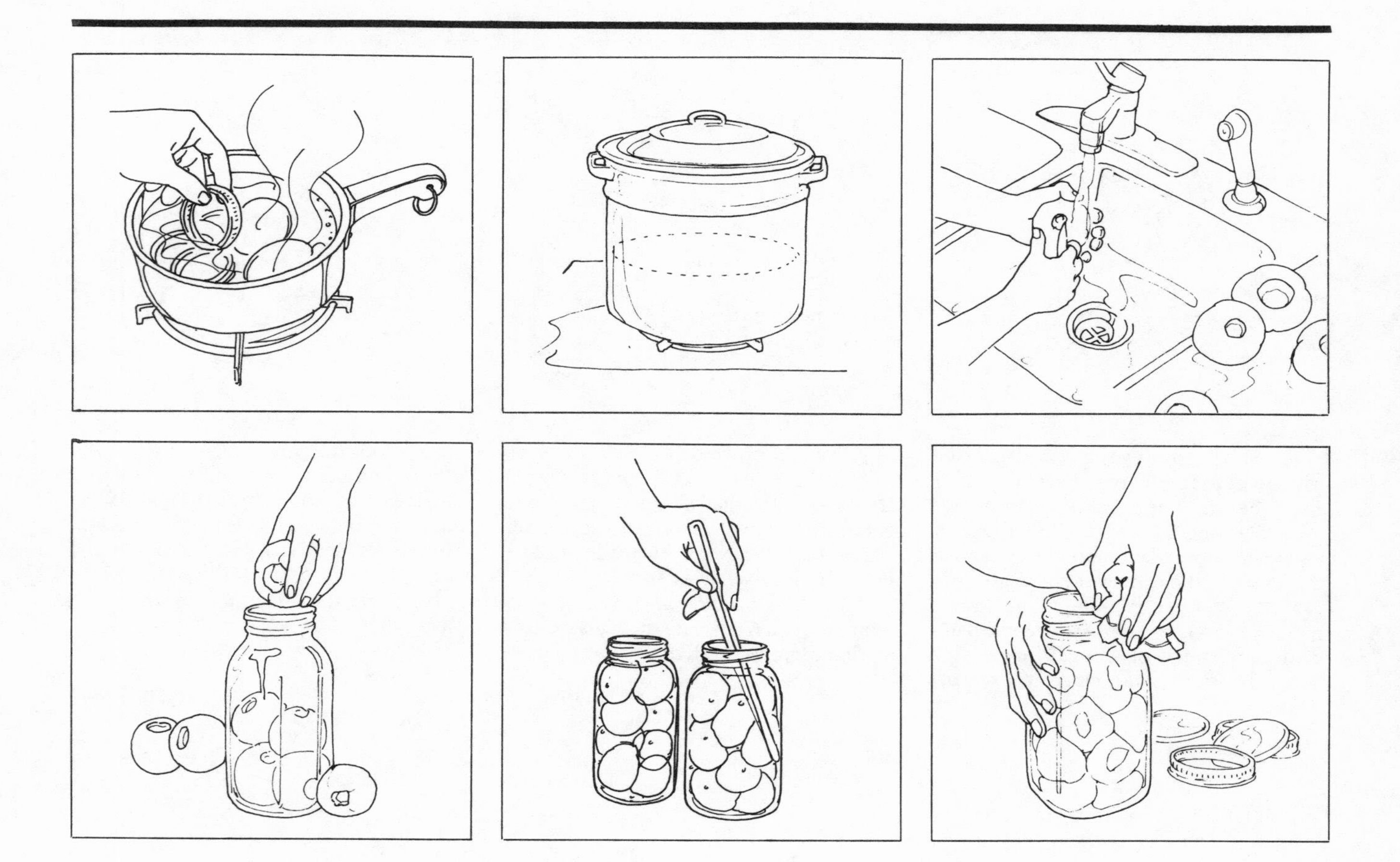

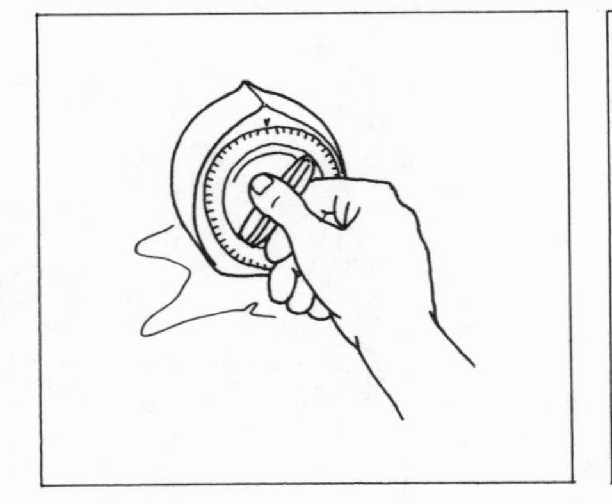

10. Start counting the processing time as soon as the water returns to a full boil. Keep the water at a full boil throughout the processing time. (A tea kettle of boiling water should be on hand to replenish the water in the canner if necessary.)

11. If you are canning more than 1 batch of food, prepare the second batch while the first is processing. While the last batch is processing, clean your work area.

12. When processing time is up, carefully remove the cover from the canner. Using long-handled tongs or jar lifters, *carefully* remove jars from the canner. Place the jars several inches apart on a heavy bath towel away from drafts.

If the screw bands are loose, do not tighten them. Complete the seals on glass or zinc caps, as needed.

13. Cool the jars for 24 hours. After the cooling period is completed, remove the screw bands and check the seals. Wipe the sealed jars with a clean, damp cloth. Then label clearly with the product and date.

14. Store the jars in a cool, dark, dry place where they will not freeze.

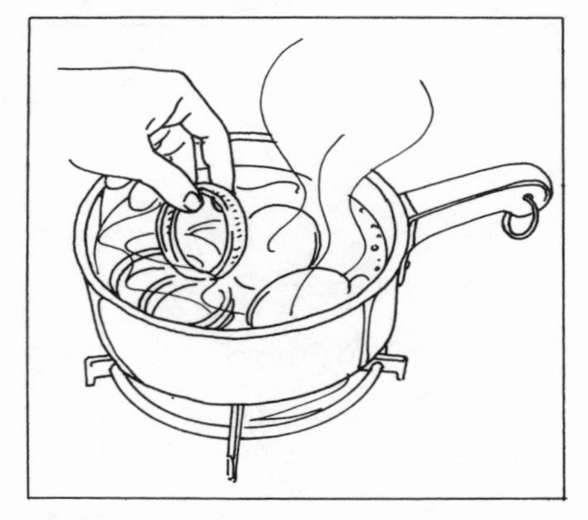

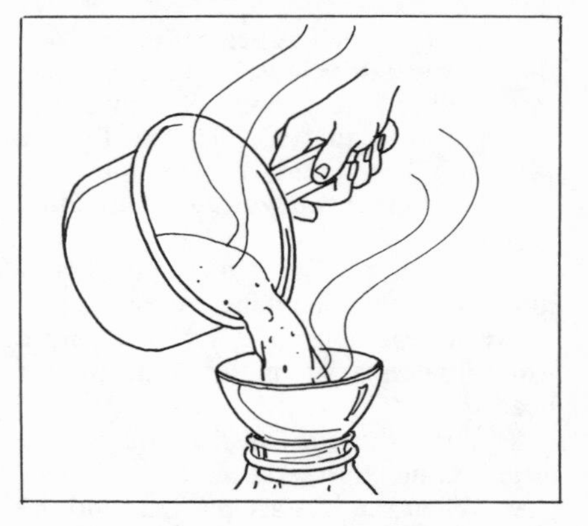

Pressure Canning Low-Acid Vegetables: 17 Steps

Pressure canning is a method of food processing using pressurized steam. It is the *only* safe method that can be used with low-acid foods.

1. Make sure your work area and all equipment are spotlessly clean. If your jars have not been prewashed, wash and rinse them. Check for nicks and cracks. If you plan to hot pack, keep the jars hot in water.

Assemble your equipment and set your tools where they will be most convenient. You will need a scrub brush, colander and/or strainer, paring and chopping knives, cutting board, food processing equipment, measuring cups and spoons, bubble expeller (wooden chopsticks work well), wide-mouth funnel, large bowls, canning jars, pressure canner, preserving kettle, tea kettle, canning lids and screw bands in hot water, large wooden and slotted spoons, soup ladle, jar lifter or tongs, timer, hot pads, mitts, or heavy potholders, and heavy bath towels.

2. Place jar lids and screw bands in water according to the manufacturer's directions. Place glass jar lids and rubber jar rings or zinc caps and rubber jar rings in hot water.

3. Select perfect fresh vegetables at the peak of their maturity. Wash the vegetables. Drain.

4. Prepare the vegetables: peel, slice, dice, julienne, puree, or leave whole.

5. Begin heating 2 quarts of water in the tea kettle for filling jars and 2 quarts of water in the canner.

6. Pack the jars.

Cold pack. Firmly pack clean, unheated jars with vegetables. If desired, add ½ teaspoon salt to pints, 1 teaspoon to quarts. Fill the jars with hot, not boiling, water or vegetable juice, leaving proper head space, according to instructions with each individual vegetable.

Hot pack. Place clean, prepared vegetables in a large pan and cover with water or juice. Bring to a boil and simmer for 5 minutes for dense vegetables, such as whole carrots or beans, and 3 minutes for smaller ones, such as peas or corn. Pack the hot vegetables loosely into clean, hot jars, adding ½ teaspoon salt to pints and 1 teaspoon to quarts, if desired. Fill the jars with hot liquid, leaving the head space recommended with each individual vegetable.

7. Expel air bubbles in the jar by running a nonmetallic kitchen utensil gently between vegetables and jar. Add more water if necessary to maintain the proper head space.

8. Wipe the rim of the jars with a clean, damp cloth, and seal.

9. When the jars are packed, and the lids are in place, set the jars on the rack in the canner so that steam can flow around the jars. The jars should not touch.

10. Put the cover on the canner and lock in place according to the manufacturer's directions. Heat the canner, following the manufacturer's directions for regulating the steam flow and exhausting the canner. Exhausting the canner, or venting, means allowing steam to escape from the petcock or steam valve of the pressure canner, so that air trapped in the canner can escape. If the canner is not properly exhausted, the air will cause the reading on the gauge to be inaccurate, and your canning temperatures may be too low for safety. Sometimes this also prevents a good seal. To properly exhaust

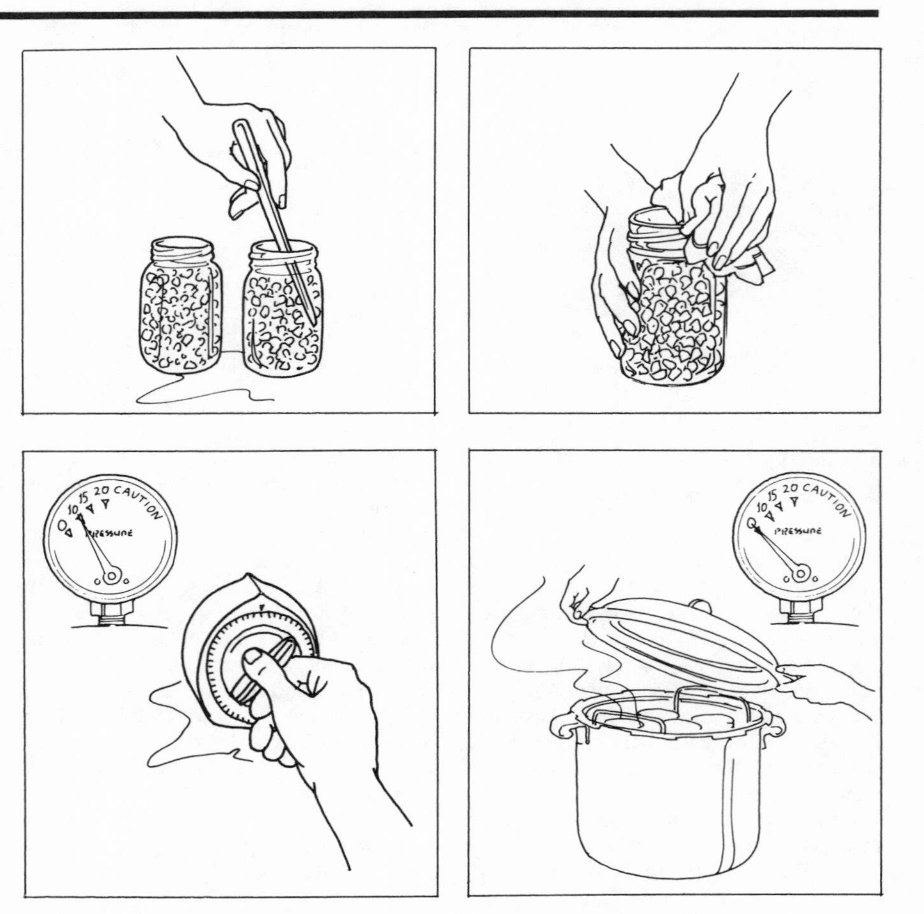

the canner, you should see a steady stream of steam escape for 7–10 minutes; check your manufacturer's directions. Close the vent on canners that have gauges after exhausting.

11. Start timing when your canner has reached the required pressure (usually 10 pounds of pressure, or adjusted for higher altitudes: see p. 127). Process for the exact length of time given under the individual vegetable or recipe. *Make sure to keep the pressure constant. If the pressure falls below the required amount, you must start counting the processing time all over again.*

12. If you are canning more than 1 batch, prepare the second batch while the first is processing. While the last batch is processing, clean up your work area.

13. Turn off the heat when the processing time is up. Remove the canner from the heating element of your range if it is electric. Let the canner stand *unopened*, until the pressure is reduced to zero. Do not try to hurry this step; it is very important for the pressure to go down slowly. This may take up to 45 minutes.

To tell if canners with weighted controls are cooled sufficiently, nudge the control with a pencil; if you do not *see* any steam, the pressure is down. (You will always hear a hissing sound; you must *see* the steam.) A dial gauge will show the pressure at zero when the jars are ready to be removed.

14. Remove the weight control (if you have that type of canner), and unlock the cover. Always tilt the covers of pans containing hot liquids or steam as you open them to direct the steam away from you. The steam should come out of the pan on the far side.

Allow the jars to cool in the canner for 10 minutes.

15. Using long-handled tongs or a jar lifter, *carefully* remove jars from the canner. Place the jars several inches apart on a heavy bath towel away from drafts. Do not cover the jars while they cool. If the screw bands are loosened, do not tighten. Complete seals with glass or zinc lids, if needed.

16. Cool jars for 24 hours. Then remove screw bands and check the seals. Wipe sealed jars with a clean, damp cloth. Label clearly with the product and date.

17. Store the jars in a cool, dark, dry place.

If you do not have a dishwasher, a large kettle of hot, sudsy water will soak dishes out of the way of the work area, making clean-up chores easier when the sink is free.

CHAPTER 4

Preserving Each Vegetable

There is a *best* way to preserve each vegetable; there is a *fastest* way as well—and sometimes the best way is the fastest way, but not always. The best way means that the vegetable will taste closest to fresh when you cook it for the table.

This summer, I froze, canned, pickled, and stored every vegetable in this chapter. I also timed my work. What I want to show is how much time each method took me, so you can plan accordingly.

Suppose your beans are ready to harvest. Turn to the page on beans (the vegetables are listed in alphabetical order) in this chapter. Decide which preservation method to use. Look at the timing. Do you have the time to freeze or pressure can those beans today? Fine. Then read the harvest tips for beans, and go into the garden and pick the beans.

What if you don't have enough time to process the beans, you ask? Those beans are going to get too mature out in the garden. You don't want to wait to pick them.

In chapter 2, I shared some of the different shortcuts I have used in my kitchen—such as storing fresh vegetables overnight in garbage cans filled with ice, or tray freezing unblanched beans to be used later in relishes. Check the index or skim through chapter 2 to find shortcuts that will work for you.

Please Review the Basic Techniques. Before you start canning or freezing, turn back to chapter 3 for a quick review of the basic techniques. Assembling all your equipment, setting up a productive work flow, and ensuring the safety of your methods will help to save you time in the long run.

With the basic techniques fresh in your mind, follow the step-by-step instructions I have provided for each vegetable.

Timings Will Vary

The timings given for individual vegetables are approximate. The time for preparation—washing, chopping, pureeing, slicing, and so on—will vary, depending on the food-processing equipment you are using, your experience, and how fast or slow a worker you are.

My times are given to show you how quickly each process can be accomplished with experience (it comes with time, believe me), the proper equipment, and efficient use of your time.

Pressure canners will vary in come-up time, depending on the size of the canner. Timings given in this book are for the large 22-quart canner.

Timings will also vary if you use different equipment than the ones specified.

Ready to harvest? Let's start with beans.

Beans: Green and Yellow Wax

The best beans for preserving are those that are slightly immature. Green beans should be long and slender and have tiny seeds. Yellow wax beans should still have a little green tint to their flesh. Harvest beans late in the day. The foliage should be dry. Harvesting beans from wet plants may cause the developing crop to rust.

Beans can be left to dry on the vine past maturity. After the pods are dry, shell them. Dry the beans on cookie sheets for several days. Place the beans in airtight containers and store in a dry place. These beans may be used as regular dried beans in soups and casseroles.

Can't preserve right away? Do not wash beans. Store by one of the methods suggested in chapter 2 (pp. 12–13).

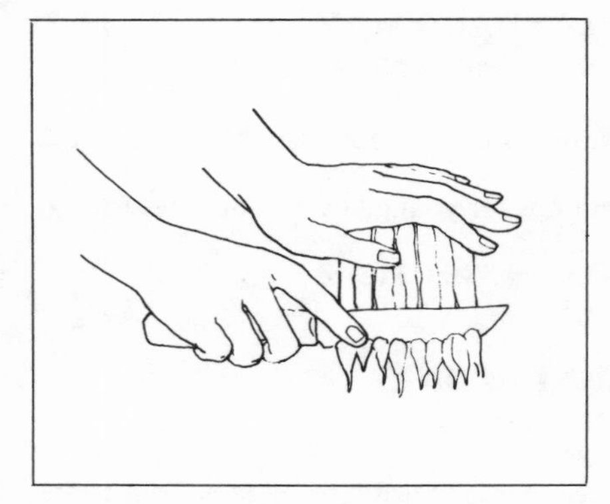

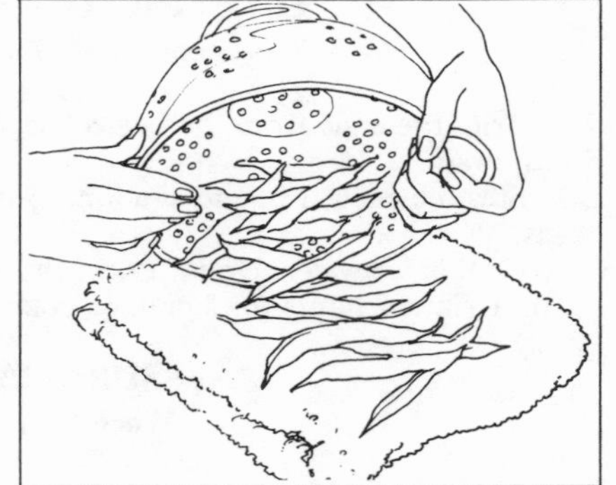

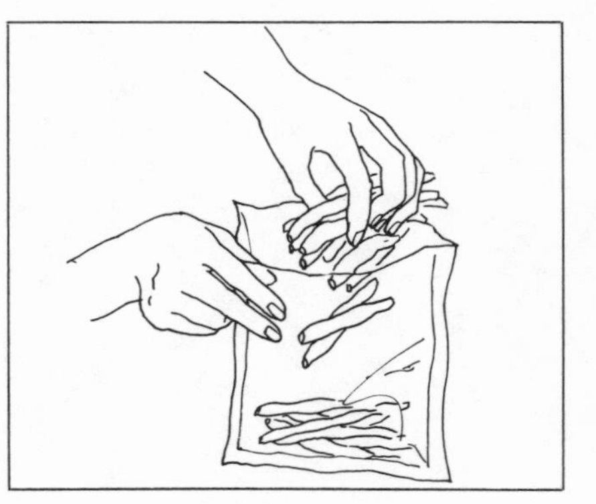

FREEZING UNBLANCHED WHOLE OR JULIENNE BEANS

Fastest, Best Finished Product

1. Trim ends of beans.
2. Wash, drain, pat thoroughly dry. Julienne, if desired.
3. Pack in gallon-size freezer bags. Press out air. Seal. Freeze.

Total Time: 45–65 minutes for 15 pints (15 pounds)

Note: Unblanched beans will retain good quality for 6 months.

Frozen beans should be cooked by steaming or stir-frying for best flavor and texture, or cooked in the boilable bag, for 18–20 minutes. The less contact beans have with water during freezing and cooking, the better flavor and texture they will have.

FREEZING IN BOILABLE BAGS

Fast, Easy, Excellent Finished Product

1. Begin heating water for blanching. To prepare the beans, trim the ends. Wash, drain, pat dry. Pack in boilable bags.

Add butter and seasonings, if desired. Press out air. Seal bags.

2. Blanch bags, 4 at a time, in boiling water: 6 minutes for young beans, up to 8 minutes for old beans.

3. Cool. Pat bags dry. Freeze.

Total Time: 80 minutes for 8 pints (8 pounds)

FREEZING THE STANDARD WAY

More Time-consuming, Finished Product Not as Good

1. Trim the ends from the beans. Begin heating water for blanching. Wash the beans, drain.

2. Steam blanch, 1 pound at a time: 3 minutes for young beans, 4 minutes for old beans.

3. Cool in ice water. Drain. Pack. Press out air. Seal. Freeze.

Total Time: 2 hours for 8 pints (8 pounds)

PRESSURE CANNING ONLY

When You Can't Freeze

1. Preheat 2 quarts of water in canner and 2 quarts of water to fill jars. To prepare the beans, trim the ends. Wash. Drain. Cut or julienne, if desired.

2. Pack beans in jars. Add boiling water to cover. Leave ½ inch head space. Add salt (½ teaspoon per pint, 1 teaspoon per quart), if desired.

3. Load pressure canner. Add boiling water according to manufacturer's directions.

4. Exhaust canner according to the manufacturer's directions, usually 7–10 minutes.

5. Process at 10 pounds: 20 minutes for pints, 25 minutes for quarts.

6. Allow canner to cool. Complete seals on jars, if necessary.

Total Time: 3¼ hours for 9 quarts (14–16 pounds)

Beets

Beets should be left in the garden until late fall. A few frosts will not harm them. Dig beets on a sunny day. Cut the tops off, leaving an inch of stem. Do not cut off roots. Let the beets lie on the ground until the following day. Beets stored in a root cellar will remain fresh and sweet-tasting well into late spring.

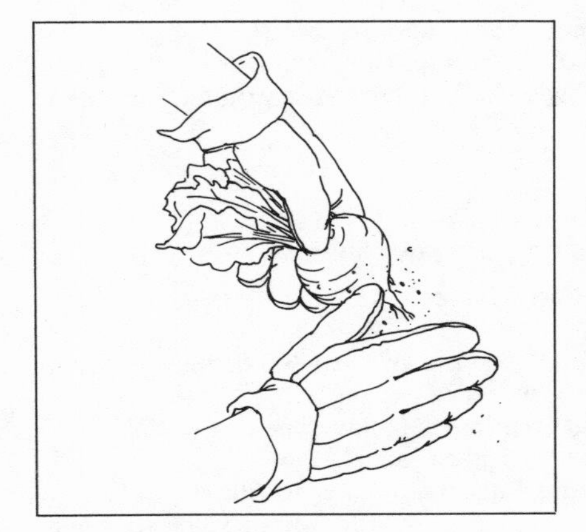

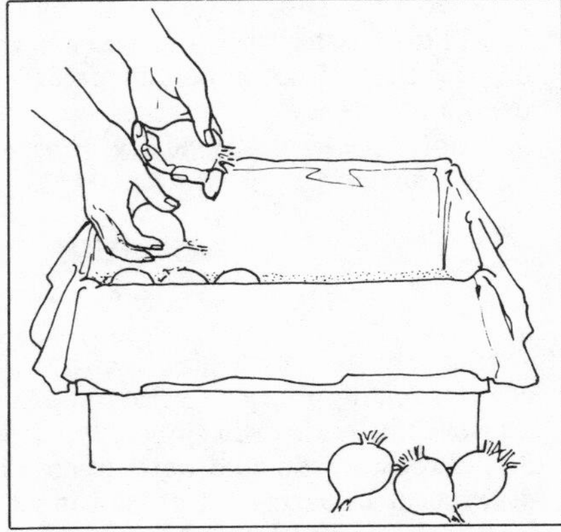

COLD STORAGE

Fastest, Best Finished Product

1. Brush off the excess soil on the roots, but do not wash. Place a large plastic bag in a cardboard box; add 2–4 inches of *fresh* sawdust (2 inches for storage areas that will remain above freezing).

2. Add a single layer of beets, leaving a 2-inch space all around the side of the layer to be filled with sawdust. Cover with a 1-inch layer of sawdust. Continue to layer the beets until the box is full. Finish with a 2–4 inch layer of sawdust.

3. Fold over the top of the bag and close the box. Store in the coldest area of the root cellar or in an unheated area, such as a garage.

Total Time: 20 minutes for 1 bushel (52 pounds)

Another method of peeling beets is to cut the tops from beets leaving 2 inches of stem to prevent bleeding. Do not cut the roots off. Scrub thoroughly. Place in a large kettle. Cover with hot tap water. Bring to a boil and boil for 15–20 minutes for canned beets, or until tender for frozen beets. Slip off skins. The disadvantage of this method is that beets should all be of the same size; otherwise, smaller beets will get overcooked and mushy before larger ones are cooked. This does not happen when beets are baked in the oven; beets of several sizes may be baked together and none overcook.

FREEZING SLICES IN BOILABLE BAGS

Excellent Finished Product

1. Cut tops and roots off close to the beet. Preheat oven to 400 degrees F. Scrub beets thoroughly. Place on a rack in a large roaster. Cover.
2. Bake until tender (approximately 1 hour for 2½–3-inch beets).
3. Fill the roaster with cold water to cool beets. Slip off skins.
4. Slice beets. Pack in boilable freezer bags. Add butter and seasoning, if desired. Press out air. Seal.
5. Cool in ice water. Wipe dry. Freeze.

Total Time: 2¼ hours for 8 pints (12 pounds)

PRESSURE CANNING ONLY

When You Can't Freeze

1. Cut tops and roots off close to the beet. Preheat oven to 400 degrees F. Scrub beets thoroughly. Place on a rack in a large roaster. Cover.
2. Bake until skins will slip easily, about 30 minutes for average-size beets.
3. Fill roaster with cold water to cool beets. Slip off skins.
4. Preheat 2 quarts of water in canner and 2 quarts of water to fill jars.
5. Slice, dice, or leave small beets whole. Pack in jars. Add boiling water to cover. Leave ½ inch head space. Add salt (½ teaspoon per pint, 1 teaspoon per quart), if desired.
6. Load canner and exhaust according to manufacturer's directions, usually 7–10 minutes.
7. Process at 10 pounds: 30 minutes for pints, 40 minutes for quarts.
8. Allow canner to cool. Complete seals on jars, if necessary.

Total Time: Approximately 4½ hours for 7 quarts (21 pounds)

When blanching broccoli, make long cross cuts in the stalk, so that the stalk will cook in the same length of time as the bud.

Broccoli

Harvest broccoli when the head stops growing, but before the individual clusters start to spread out. Second crop clusters (side shoots) should be cut while the buds are still clustered tightly.

Can't preserve right away? Do not wash. Store by one of the methods suggested in chapter 2.

FREEZING UNBLANCHED BROCCOLI

Fastest Method, Good Finished Product

1. Soak broccoli in cold salted water for ½ hour to remove dirt and insects. Rinse well. Drain. Divide into uniform-size pieces or chop.

2. Pack in gallon-size freezer bags. Press out air. Seal. Freeze.

Total Time: 45 minutes for 4 pounds (4 pints)

For best flavor and texture, frozen broccoli should be steamed or stir-fried.

FREEZING THE STANDARD WAY

Best Finished Product

Unblanched broccoli florets retain good quality in the freezer for only 6 weeks, while the stalks hold up for 3 months. Save time by freezing sliced pieces of stalks without blanching, and blanch the florets for the best finished product. The sliced stalks make an excellent addition to mixed stir-fried vegetables.

1. Soak broccoli in cold salted water for ½ hour to remove dirt and insects. Meanwhile, preheat water for steam blanching.
2. Drain broccoli. Rinse. Divide into uniform-size pieces or chop.
3. Steam blanch, 1 pound at a time for 5 minutes. Then, cool, drain, pack, press out air, and seal. Freeze.

Total Time: 1¾ hours for 8 pints (8 pounds)

Coarsely chopped, cabbage can be frozen unblanched to be used in soups or casserole dishes within 4–6 months. Steam-cook frozen cabbage for casserole dishes requiring cooked cabbage.

Cabbage

Cabbage for winter keeping should be left in the garden as late in the fall as possible. Pick just before the first frost. If the heads indicate a tendency to split, and it's too early to harvest, give the heads a quarter turn to break off some of the roots; this will slow down the growth and can be repeated every 7–10 days. When ready to harvest, pull the entire plant up from the garden, roots and all.

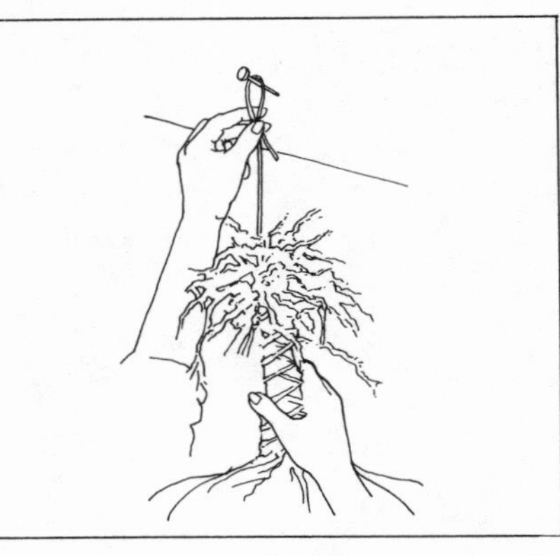

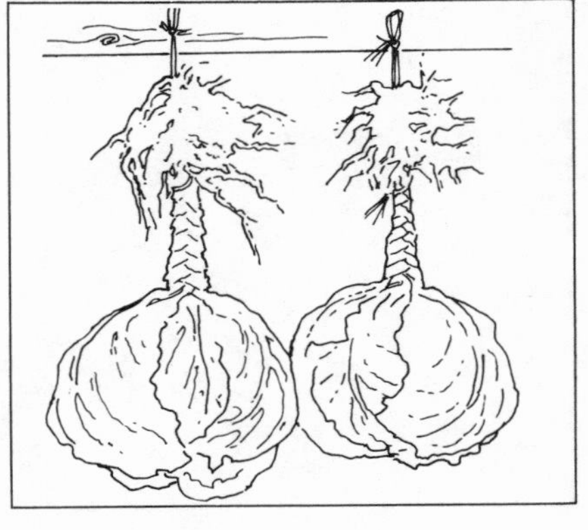

COLD STORAGE
Fastest Method, Best Finished Product

1. Pull up cabbage plants with roots intact.
2. Transfer freshly pulled cabbage plants directly to the root cellar, and hang upside down on hooks or nails.

Finely shredded cabbage can be made into freezer coleslaw (p. 91).

3. Leave the outer leaves on the plant; they will form a dry paperlike covering that will help to keep the cabbage fresh.

Cabbage can be brined as sauerkraut or made into relishes; for recipes, see chapter 5.

Outer leaves of cabbage can be frozen whole and unblanched to be used as wrappers for baked stuffed cabbage leaves. Since the leaves will wilt when defrosted, it will not be necessary to drop them into boiling water before preparing this dish.

Carrots

Carrots should be left in the garden until you are ready to store or process them. Carrots for cold storage are best if harvested late in the fall, though they should not be allowed to become oversized and stringy. A few light frosts will not harm them. Carrots stored in a root cellar will remain fresh and sweet-tasting well into late spring.

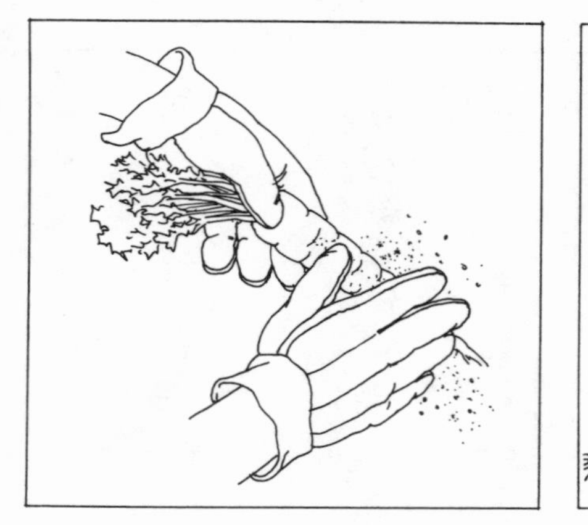

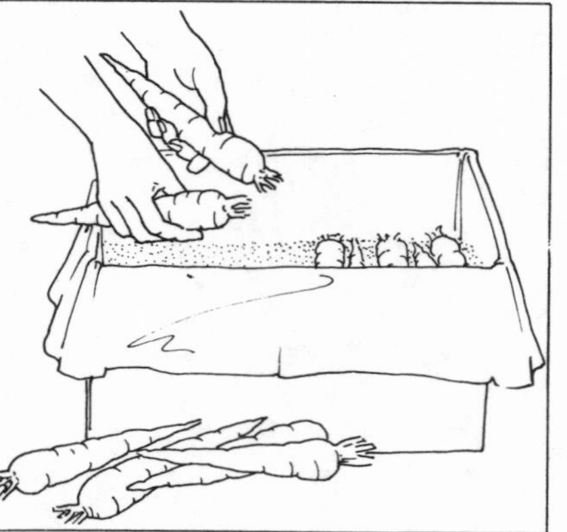

COLD STORAGE

Fastest Method, Best Finished Product

1. Dig carrots on a sunny day; cut off tops close to the carrot and let them lie on the ground all day or even overnight to kill the little feeder roots. Do not wash the carrots.

2. Brush off excess soil. Place a large plastic bag in a cardboard box; add 2–4 inches of *fresh* sawdust (2 inches for storage areas that will remain above freezing). Add a single layer of carrots, leaving a 2-inch space all around the sides of the box to be filled with sawdust. Cover with a 1-inch layer of sawdust. Continue until box is full, finishing with a 2–4-inch layer of sawdust.

3. Fold over top of the bag and close the box. Store in the coldest area of the root cellar or an unheated area such as a garage.

Total Time: 20 minutes (does not include harvesting) for 50 pounds

Mature fall carrots give a better finished frozen product than young early carrots. They should be peeled before freezing.

Carrots can be left in the ground, covered with a heavy mulch, and dug up throughout the winter or early in the spring; however, because of melting snows and freezes, I have found this to be an unsatisfactory method of storage in my area of northwestern Vermont.

FREEZING IN BOILABLE BAGS
Excellent Finished Product

1. Begin heating water for blanching. Scrub carrots thoroughly; peel. Slice or julienne with food processor.
2. Pack in boilable bags. Add butter and seasonings, if desired. Press out air. Seal.
3. Blanch bags (4 at a time) in boiling water, for 8–10 minutes.
4. Cool. Pat bags dry. Freeze.

Total Time: 1⅔ hours for 8 pints (12 pounds)

FREEZING THE STANDARD WAY
Good Finished Product

1. Scrub carrots thoroughly; peel. Begin heating water for blanching. Slice or julienne carrots with food processor. Whole carrots do not freeze well.
2. Blanch, 1 pound at a time, by steam or immersion, for 3–4 minutes.
3. Cool. Drain well. Pack. Press out air. Seal. Freeze.

Total Time: 1¾ hours for 8 pints (12 pounds)

PRESSURE CANNING ONLY
When You Can't Freeze

1. Scrub carrots, peel. Preheat 2 quarts of water in canner and 2 quarts of water to fill jars. Slice or julienne carrots with a food processor or leave whole.
2. Pack into jars. Add boiling water to cover. Leave 1 inch head space. Add salt (½ teaspoon per pint, 1 teaspoon per quart), if desired.
3. Fill canner according to manufacturer's directions.
4. Load canner and exhaust according to manufacturer's directions, usually 7–10 minutes.
5. Process at 10 pounds: 25 minutes for pints, 30 minutes for quarts.
6. Allow canner to cool. Complete seals on jars, if necessary.

Total Time: 4½ hours for 7 quarts (21 pounds)

Corn

Harvest your corn as soon as the kernels become full and sweet. Corn is at the right stage for picking when you can press milky fluid from the kernels. If the fluid is too clear, the corn is not ready; if it is thick, the corn has gone past its prime and will taste tough and starchy. Starchy corn is best used in cream-style corn.

Some of the newer hybrid super sweet varieties will retain their prime flavor up to 4 days, if left right on the stalk. Others should be picked as soon as they are ready. Be sure you know which type you planted.

If you must harvest corn before you are ready to process it, harvest it late in the afternoon and do not husk. Place the corn in a large barrel with a layer of ice between each layer of ears. Cover with several layers of newspaper and a cover, if possible. Corn will keep this way for about 24 hours, before losing its sweet flavor.

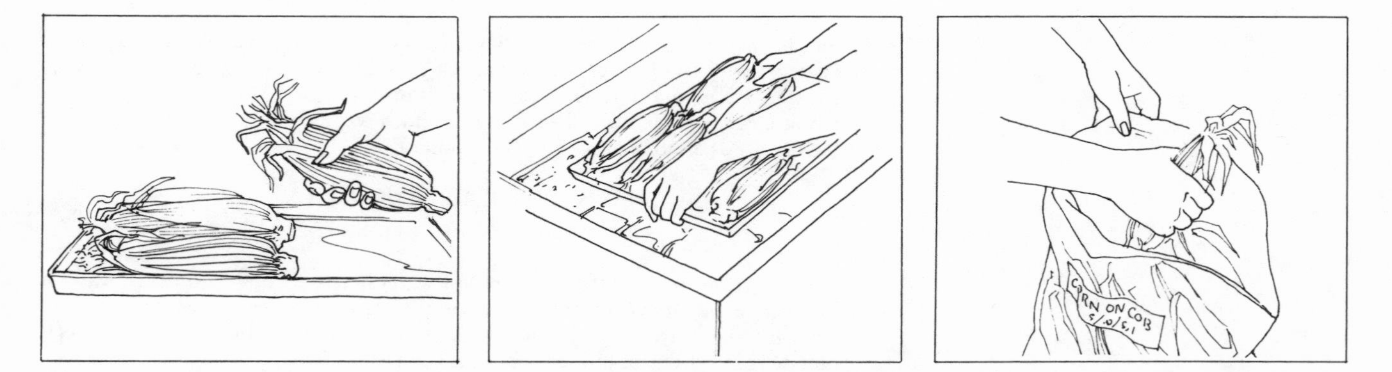

FREEZING UNBLANCHED CORN IN HUSKS

Fast and Easy

1. Harvest the corn, and do not husk. Place the ears on cookie sheets, or just put them in the freezer loose in a single layer. Freeze for 48 hours.

2. Bag the corn in a large plastic food-safe bag.

3. Return to freezer. Remove ears as needed. Eat within 4 months.

To cook corn frozen in husks, husk the frozen corn under cold running water. Remove the silks by rotating your hands around the corn. Put the corn in a pan, cover with cold water. Cover the pan, heat to boiling, and boil for 1 minute. Remove from heat, and let the corn stand in hot water 5 minutes. Eat immediately.

Corn on the cob frozen in boilable bags should be cooked by placing the bags in cool water. Bring the water to a boil and cook the corn for 10 minutes, turning the package once halfway through the cooking time. Remove from boiling water immediately. Bags of cut corn should be placed in cool water. Bring the water to a boil, and cook for 5 minutes.

FREEZING CORN ON THE COB IN BOILABLE BAGS

Fast, Best Finished Product

1. Begin heating water for blanching. Husk cobs, trim, and pack in boilable bags. Add butter if desired. Press out air. Seal.
2. Blanch bags, 4 at a time, in boiling water, for 10 minutes. After 5 minutes, turn bags over.
3. Cool in ice water. Pat bags dry. Freeze.

Total Time: 55 minutes for 8 packages of 4 ears

FREEZING CORN ON THE COB THE STANDARD WAY

Good Finished Product

1. Husk cobs. Begin heating water for blanching. Trim.
2. Blanch by steam or immersion, 6 ears at a time, for 6–8 minutes.
3. Cool. Drain. Pack in bags. Press out air. Seal. Freeze.

Total Time: 60 minutes for 32 ears

FREEZING CREAM-STYLE CORN

Excellent Finished Product

1. Husk. Cut corn from cob by cutting down through the center of the kernel. Scrape remaining corn from the cob.
2. Measure kernels as you pour into a large kettle. Add ¼ cup boiling water and 1 teaspoon cornstarch per cup. Stir well. Bring to a boil, simmer for 5 minutes, stirring frequently. Fill sink with ice water while cooking corn.
3. Place kettle in sink of ice water. Stir to cool.
4. Package in rigid containers or bags and freeze.

Total Time: 1½ hours for 8 pints (32 large ears)

To cook frozen cream-style corn, place the frozen corn in a double boiler. Add a small amount of milk or light cream, salt, pepper, and a pat of butter. Heat until just piping hot. Let it stand 2–3 minutes before serving.

WHOLE KERNEL CORN IN BOILABLE BAGS

Fast Method, Best Finished Product

1. Husk. Begin heating water for blanching. Cut corn from cob. Pack in boilable bags. Add butter if desired. Press out air. Seal bags.
2. Blanch 4 bags at a time in boiling water for 6 minutes.
3. Cool. Pat bags dry. Freeze.

Total Time: 60 minutes for 32 ears

PRESSURE CANNING WHOLE KERNELS

When You Can't Freeze

1. Husk, cut kernels from cob. Begin heating 2 quarts of water in canner and 2 quarts of water to fill jars.
2. Pack kernels into pint jars, add boiling water to cover. Leave 1 inch head space. Add salt (½ teaspoon per pint), if desired.
3. Load pressure canner. Add boiling water according to manufacturer's directions.
4. Exhaust canner according to manufacturer's directions, usually 7–10 minutes.
5. Process pints at 10 pounds for 55 minutes.
6. Allow canner to cool. Complete seals on jars, if necessary.

Total Time: 3¾ hours for 7 pints (28 ears)

PRESSURE CANNING CREAM-STYLE CORN

1. Husk. Cut corn from cob by cutting down through the center of the kernel. Scrape the remaining corn from the cob. Heat 2 quarts of water in canner and 2 quarts to fill jars.
2. Pack in pint jars. Add boiling water to cover. Leave 1 inch head space. Add ½ teaspoon salt per jar, if desired.
3. Load pressure canner. Add boiling water according to manufacturer's directions.
4. Exhaust canner according to manufacturer's directions, usually 7–10 minutes.
5. Process pints at 10 pounds for 85 minutes.
6. Allow canner to cool.

Total Time: 4¼ hours for 7 pints (28 ears)

Cucumbers

Cucumbers can be picked at any stage: tiny, for whole pickles; medium, for dill, sliced, or chunk pickles; large, for ripe cucumber pickles. For eating fresh, cucumbers should be harvested when they are 4–6 inches long and about 1–1½ inches in diameter.

Cucumbers cannot be canned or frozen by any of the regular methods; they must be preserved in brine or vinegar or frozen for salads. Thinly sliced and frozen in brine, they produce a delightfully fresh, crisp flavor to be used in salads all winter.

When cucumbers have reached just the right stage for a specific type of pickle, they should be picked immediately. Even 1 day can allow too much growth. Do not wash cucumbers; store by one of the methods suggested in chapter 2.

FREEZING SLICES FOR SALADS

Fastest Method, Excellent Product

1. Wash 6–8 firm, slender cucumbers. Slice thinly with a food processor. Peel and thinly slice 1 medium onion. Measure out 2 quarts of sliced cucumbers. Mix with onions and 2 tablespoons of salt in a large bowl.
2. Let stand for 2 hours.
3. Drain vegetables well. Rinse thoroughly with cold water. Drain well again. Return drained vegetables to the rinsed bowl. Add ⅔ cup each of oil, vinegar, and sugar. Add 1 teaspoon celery seed. Mix well.
4. Cover and refrigerate overnight.
5. Pack in freezer jars or rigid freezer containers. Leave 1 inch head space. Freeze.

Total Time: 1 hour (does not include standing time) for 4 pints

Note: Can be defrosted and eaten after 1 week. Store defrosted cucumbers in refrigerator. Keeps well after defrosting for several weeks.

SALT
SUGAR
VINEG

Greens

Spinach to be preserved should be harvested when the leafy portion is 6 inches long. Chard and other greens should be harvested while still young and tender; chard can be as long as 10 inches. Spinach bolts and turns bitter quickly when the weather warms up, so watch it carefully. Greens should be harvested early in the day while there is still dew on the leaves and before the warming rays of the sun cause them to go limp. If cut at this point they will stay fresh as long as 2–3 days.

Can't preserve right away? Do not wash greens. Store by one of the methods suggested in chapter 2. Avoid placing heavy packages of ice on top of the tender leaves.

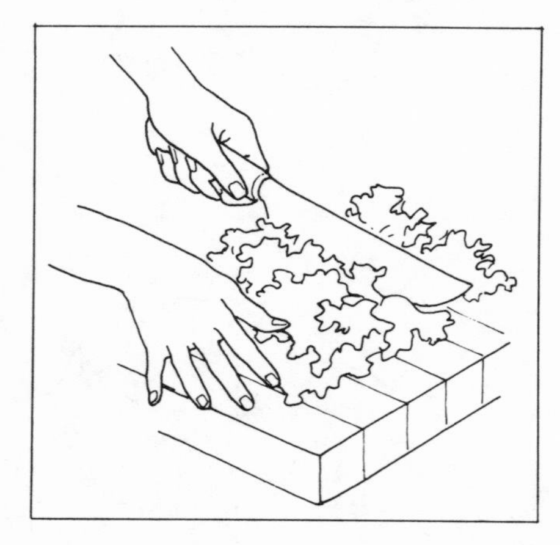

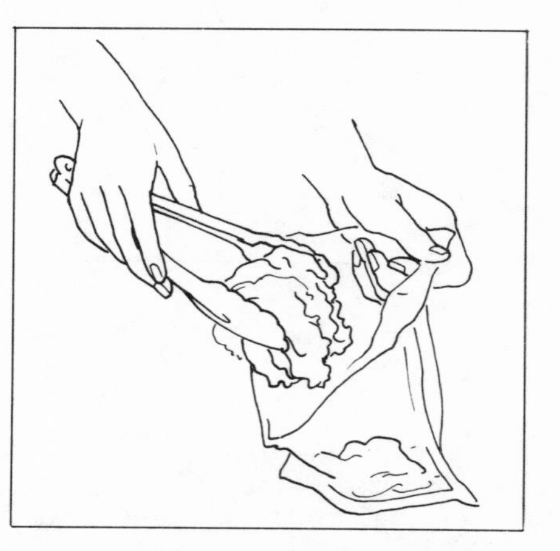

STIR-WILTED GREENS

Fastest Method, Best Finished Product

1. Wash, pick over, drain. Cut through with large knife.

2. Stir-fry until wilted: 2–3 minutes.
3. Pack in boilable bags.

4. Cool, pat bags dry, freeze.

Total Time: 1½ hours for 8 pints (12 pounds)

Greens can be frozen by using the standard method of freezing. That method takes more time and since the product is not as good, I do not recommend it.

PRESSURE CANNING ONLY
When You Can't Freeze

1. Wash; pick over carefully. Drain. Preheat 2 quarts of water and jars in canner, and preheat 2 quarts of water to fill jars. Preheat water in steamer. Process as many greens as the steamer will hold. Steam until wilted.
2. Pack hot greens in heated jars. Add boiling water to leave ½ inch head space. Add salt (½ teaspoon for pints, 1 teaspoon for quarts), if desired.
3. Load canner and exhaust according to manufacturer's directions, usually 7–10 minutes.
4. Process at 10 pounds: 70 minutes for pints, 90 minutes for quarts.
5. Allow canner to cool. Complete seals on jars if necessary.

Total Time: 4¼ hours for 8 pints (10–12 pounds)

Lettuce

Late-planted fall lettuce can be harvested right up until the first frost.

EXTENDING THE HARVEST

1. To preserve lettuce for winter use, plant seed in the garden in late summer.
2. Transplant to a cold frame just before the first frost.
3. Properly cared for, lettuce will continue to produce well into the early winter. See p. 128 for a list of reading material on extending the harvest season.

Freeze a mixture of onions and green or moderately hot peppers to be fried and eaten with hamburgers or frankfurters.

Onions, Garlic, and Shallots

Onions, garlic, and shallots, all members of the same family, are the easiest vegetables to harvest and store. Wait until their tops have fallen over in the late summer or early fall and the stems have almost completely dried. Do not try to hasten maturity by bending over the tops before they are ready to fall themselves.

Pull the onions up and allow them to dry in the sun. The longer they are dried before storage, the better they will keep. As a rule of thumb, I never put my onions in storage until the outer skins are totally dry and slip off easily—usually 2–3 weeks. I dry my onions on an old screen door that has been placed up across sawhorses. This speeds the drying process and prevents rot that could be caused by a week of constant rains.

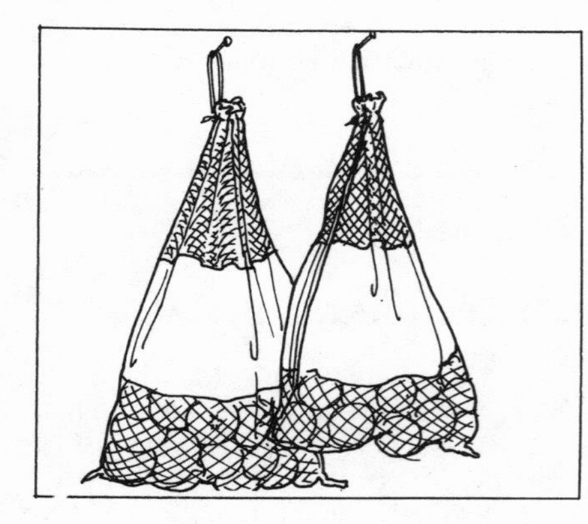

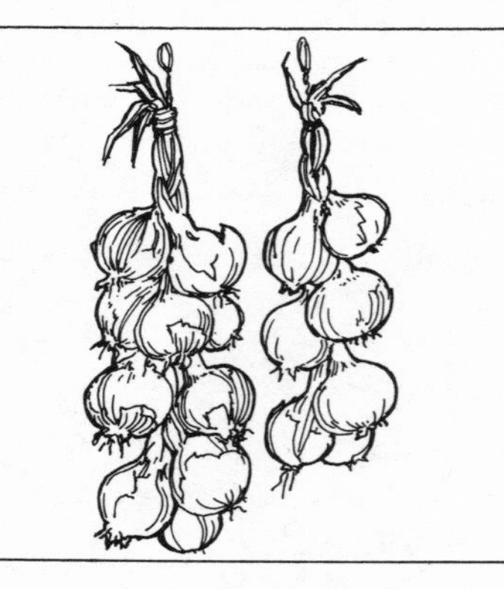

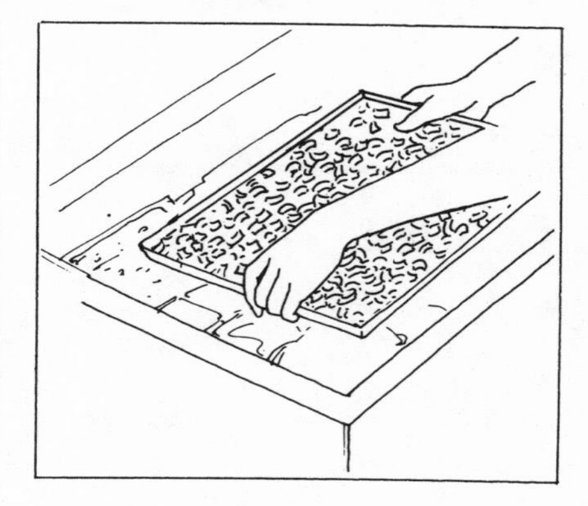

COLD STORAGE

The fastest and easiest method of storing dried onions, garlic, and shallots is to pack them into mesh bags and store in a cool, dry area. The next best method is to braid them before the stems are completely dried. Finish drying out in the sun and store in a cool, dry area. It takes me approximately 15 minutes to braid a foot-long string of onions.

Onions also can be chopped or sliced and tray frozen. Once frozen, pack in plastic bags and use as needed for cooking. One pound of onions can be peeled and chopped in a food processor in 5 minutes.

Do not store onions in humid areas. They tolerate cold temperatures as low as 38 degrees F. quite well, but will spoil quickly if the humidity is high.

Peas that go past maturity can be left to dry on the vine. Shell and store in dry airtight containers to use in soups.

Peas

Green peas should be harvested as soon as the peas fill the pod. Older peas lose their sweetness quickly and become tough. Harvest from the bottom of the plant first, since these are first to mature. Pick peas late in the afternoon for sweetest flavor. Snow peas must be picked before the pea seed starts to develop. Sugar Snap peas can be picked at any stage, right up until they fill the pod tightly, to freeze for stir-frying, which is the only cooking method I recommend for this variety. If you can't preserve right away, chill your peas immediately by storing according to one of the methods suggested in chapter 2.

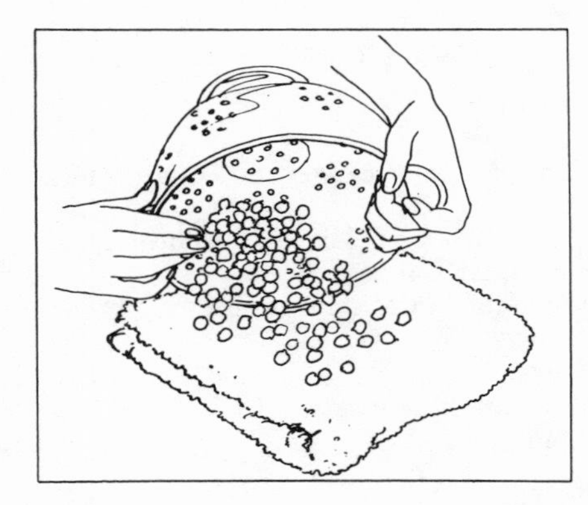

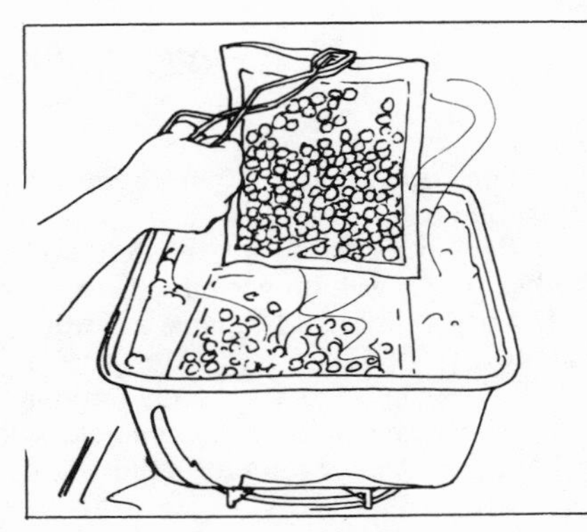

FREEZING PEAS IN A BOILABLE BAG

Fastest, Best Finished Product

1. Begin heating water for blanching; shell peas, wash, and drain. Pack in 1-pint boilable freezer bags. Add butter and seasoning, if desired. Press out air. Seal.

2. Blanch 4 bags at a time in boiling water, for 4 minutes.

3. Cool bags in ice water. Pat dry. Freeze.

Total Time: 1½ hours for 8 pints (16 pounds)

Peas packed in boilable freezer bags can be cooked in the bag in boiling water for 20 minutes. Loosely packed peas are best when steam cooked until barely tender, 5–8 minutes.

FREEZING THE STANDARD WAY
Excellent Finished Product

1. Shell peas. Preheat water for steam blanching. Wash peas. Drain.
2. Blanch, 1 pound at a time, for 2 minutes.
3. Cool peas in ice water. Drain. Pack. Press out air. Seal. Freeze.

Total Time: 2 hours for 8 pints (16 pounds)

PRESSURE CANNING ONLY
When You Can't Freeze

1. Shell peas. Wash. Drain. Preheat 2 quarts of water in canner and 2 quarts of water to fill jars.
2. Pack peas loosely in jars. Add boiling water to cover. Leave 1 inch head space. Add salt (½ teaspoon per pint, 1 teaspoon per quart), if desired.
3. Load canner and exhaust according to manufacturer's directions, usually 7–10 minutes.
4. Process pints at 10 pounds for 40 minutes.
5. Allow canner to cool. Complete seals on jars, if necessary.

Total Time: 3¾ hours for 8 pints (16 pounds)

BLANCH AND TRAY FREEZE SNOW AND SUGAR SNAP PEAS
Fastest, Best Finished Product

Frozen Sugar Snap peas can be chopped and put in tossed salads while still frozen. If they are added just 3–4 minutes before the salad is served, they will be crisp, cold, and sweet.

1. Preheat water for steam blanching. Wash. Trim ends off pods and remove strings.
2. Steam blanch, 1 pound at a time, for 3 minutes.
3. Cool. Drain. Pat dry. Tray freeze.
4. When frozen solid, pack in large freezer bags or containers.

Total Time: 2½ hours for 15 pounds (½ bushel)

Snow peas and Sugar Snap peas do not retain their crisp texture when frozen; cook by stir-frying.

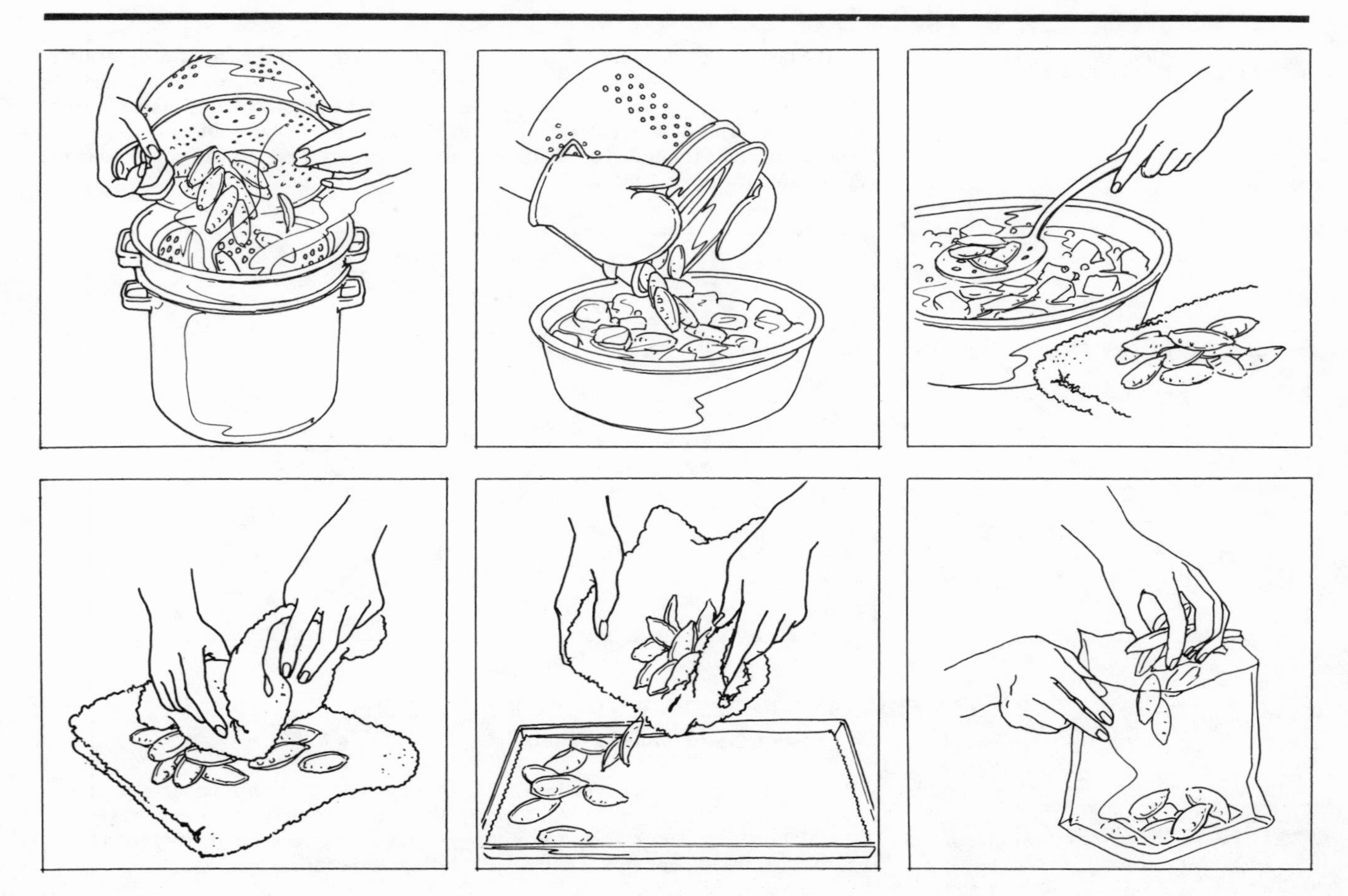

Sweet Peppers

For the best frozen product, peppers should be picked when they are fully mature in size and have thick flesh. Since peppers keep well on the bush, there is no need to pick until you are ready to use them. As peppers go past the initial stage of maturity, they will turn red. This does not affect the flavor or texture of the vegetable. I like to freeze some red peppers for added color in soups and casseroles. Peppers can be stored in the refrigerator for up to 2 weeks.

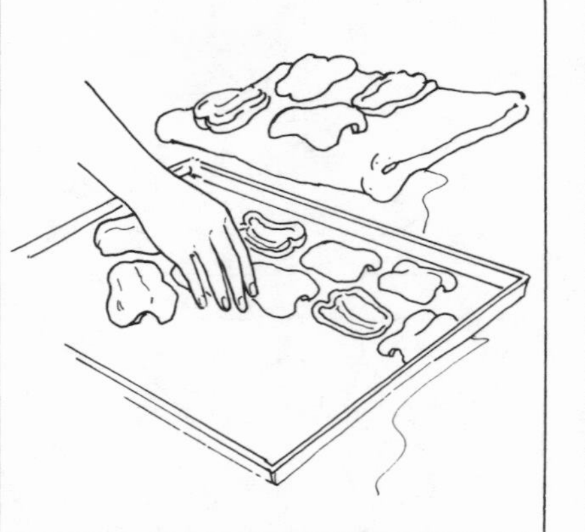

TRAY FREEZING PEPPERS WHOLE OR IN HALVES

Fastest, Best Finished Product

1. Wash; cut out stem end, seed pod, and white membrane.
2. Tray freeze whole or cut in half.
3. Package frozen peppers in 12–24 hours.

Total Time: 16 minutes for 3 pounds (approximately 12 peppers)

Note: Whole or half peppers prepared this way can be stuffed and baked; or sliced, diced, or ground from the frozen state, as needed.

Potatoes

Potatoes for winter keeping should not be dug until the tops die off. Once the tops have died, pull the tops from the ground 10 days before digging. This will prevent diseases from starting that could destroy the tubers. Do not wash potatoes before storing. After digging, store at normal temperatures (60–75 degrees F.) for 10–14 days to let the skins dry out enough to prevent moisture loss. The storage area should be protected from light as much as possible, otherwise they will develop solanine (manifested as a green coloring on the surface), and this can cause illness.

COLD STORAGE

Potatoes are easily stored in open boxes, crates, or bins in the warmer area of your root cellar (approximately 40 degrees F.) or any cool, dark room or outbuilding. Potatoes stored below 38 F. decline in starch and increase in sugar, which destroys their keeping and cooking qualities.

Do not store potatoes in the same root cellar as apples, unless both products are in well-covered containers, and the room is well ventilated.

Squash can be sliced ½ inch thick and tray frozen, unblanched, to be breaded or flour-coated and fried. Do not defrost squash before frying.

Summer Squash

Squash to be frozen or canned should be harvested when they are small and firm, and the seeds are underdeveloped. Zucchini is at its best when it is 6–10 inches long; yellow types, 5–7 inches long; and scallop types, about 4–5 inches in diameter. Squash grows fast and should not be allowed to remain on the vine past the time it reaches proper size. If you can't preserve right away, do not wash the squash. Store by one of the methods suggested in chapter 2.

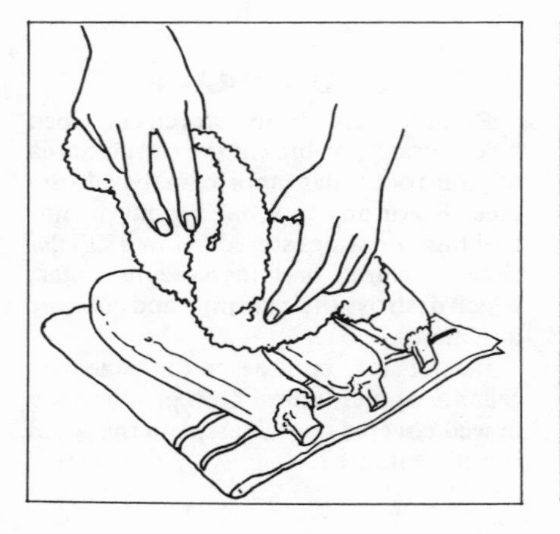

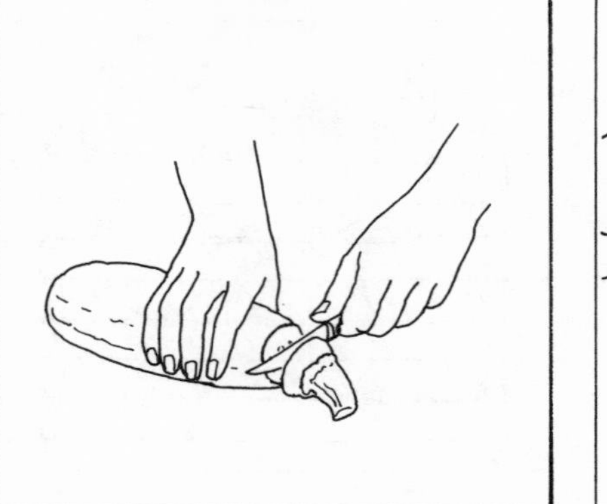

FREEZING UNBLANCHED SQUASH SLICES
Fastest, Best Finished Product for Stir-Frying

1. Wash. Drain. Pat dry.
2. Trim ends. Slice no thicker than ¼ inch.
3. Pack in gallon-size freezer bags. Press out air. Seal. Freeze.

Total Time: 10 minutes for 8 pounds

Note: Will retain good quality for 4–6 months.

Zucchini can be grated and frozen for fruit breads, cookies, and muffins. When defrosting, squeeze the moisture out of squash before measuring and adding to recipe.

For best flavor and texture, cook frozen squash by stir-frying with pork drippings or butter and seasonings (garlic is great), for 2 minutes, or steam for 4 minutes.

Summer squash can be made into pickles and relishes. (See pp. 90 and 95 for recipes.)

FREEZING SQUASH IN BOILABLE BAGS

Fast, Excellent Finished Product

1. Wash squash. Drain. Pat dry. Trim ends. Slice in ¼-inch slices. Pack in boilable bags. Add butter and seasonings, if desired. Press out air. Seal bags.
2. Blanch bags, 4 at a time, in boiling water for 5 minutes.
3. Cool bags in ice water. Pat bags dry. Freeze.

Total Time: 40 minutes for 8 pints (8 pounds)

PRESSURE CANNING ONLY

When You Can't Freeze

1. Wash. Drain. Trim ends. Slice.
2. Place the squash in a large kettle with just enough water to prevent sticking. Bring to a boil and simmer for 2 minutes. Preheat 2 quarts of water in canner and 2 quarts of water to fill jars.
3. Pack jars. Leave 1 inch head space. Add salt (½ teaspoon per pint, 1 teaspoon per quart), if desired.
4. Load pressure canner. Add boiling water according to manufacturer's directions.
5. Exhaust canner according to manufacturer's directions, usually 7–10 minutes.
6. Process at 10 pounds: 30 minutes for pints, 40 minutes for quarts.
7. Allow canner to cool. Complete seals on jars, if necessary.

Total Time: 3 hours for 7 quarts (18–20 pounds)

Tomatoes

Pick tomatoes when they are fully ripe and slightly soft to the touch. Do not allow them to become overripe, if they are to be canned in the steam canner or boiling water bath. Overripe tomatoes are too low in acid to be safely canned by these methods. You can process overripe tomatoes in the pressure canner, use them in sauces containing vinegar, or they can be frozen.

Although you will find it fastest to freeze tomatoes, canning produces a far superior end product. I recommend steam canning.

COLD STORAGE

The Very Fastest Method

At the end of the harvest season, if you would like to hasten the ripening of tomatoes, place them in a single layer in large paper bags, tied loosely. The ethylene gases they release will help them to ripen rapidly and evenly.

Tomatoes, good for fresh eating, although not as tasty as summer tomatoes, can be kept fairly well into the early winter if they are properly stored. Pick all large tomatoes just before a hard frost is predicted. Separate the green ones from those more advanced in color. Place the sorted tomatoes in a single layer in shallow boxes or on trays in a root cellar or cool garage. Cover with 2–3 layers of newspaper. Check frequently, and remove tomatoes as they ripen. Remove any tomatoes that show signs of spoilage immediately. Green tomatoes stored this way will ripen in 4–6 weeks.

FREEZING WHOLE, UNPEELED TOMATOES

Fastest Method, Least Desirable Product

To use whole unpeeled, frozen tomatoes, run them under tepid water for a few seconds to soften them slightly. Then peel and finish defrosting in a bowl. These tomatoes are best used in sauces, soups, and casseroles.

1. Wash, core tomatoes. Set tomatoes on cookie sheets and freeze.
2. When tomatoes are frozen, pack in bags.

Total Time: 30 minutes for 24 pounds

FREEZING WHOLE, PEELED TOMATOES

Good Product, Slow Method

Always defrost tomatoes packaged in plastic bags in a bowl. Use the juice that separates out, as well as the pulp, in order to keep all the nutrients.

1. Wash tomatoes. Heat water in large kettle to boiling. Refill sink with cold water. Drop tomatoes into boiling water a few at a time. Scald for 30 seconds. Remove to cold water. Lift from water, peel, and core.
2. Pack in rigid containers. Leave 1 inch head space. Freeze.

Total Time: 1 hour for 8 quarts (23–24 pounds)

FREEZING SQUEEZO-STRAINED PUREE OR JUICE

Fast Method, Best Freezer Product

1. Wash, core, and quarter tomatoes.
2. Puree through Squeezo.
3. Pack in rigid containers. Leave 1 inch head space. Freeze.

Total Time: 25 minutes for 8 pints (12 pounds)

Note: If you don't have a Squeezo or similar strainer, you can make purees and juices with a blender or food mill, but you must cook the tomatoes to soften them before straining.

Freeze or can tomatoes whole or in a puree in the summer; make up sauces, soups, chilis, catsups, relishes, and chutneys, when the woodstove is going in the winter.

STEAM CANNING TOMATO PUREE

Fastest Canning Method, Best Finished Product

1. Wash, core, and quarter tomatoes.
2. Puree in Squeezo.
3. Heat puree to boiling. Boil for 5 minutes. While the puree is heating, preheat the canner and jars.
4. Pour the hot puree into the heated jars, leaving ½ inch head space. Add salt (½ teaspoon per pint, 1 teaspoon per quart), if desired.
5. Load canner and heat on high until a steady stream of steam flows out of the canner for 5 minutes.
6. Process: pints for 10 minutes, quarts for 15 minutes.

Total Time: 1½ hours for 7 quarts (21 pounds)

Tomato juice or puree that is made in a blender or strainer must be heated to boiling and simmered for 5 minutes to remove air; otherwise the jars will not seal properly.

To use whole frozen tomatoes in salads, run them under the tepid water for a few seconds, peel, and set the tomatoes in a bowl for 30 minutes to soften slightly. Chop and serve in a bowl separate from the salad, to be added to the salad when served. After the tomatoes have been out of the freezer for an hour, they are too soft to use in salads.

STEAM CANNING COLD-PACKED WHOLE TOMATOES

Best Finished Product

1. Begin heating water in blancher, steam canner, and tea kettle. Wash tomatoes. Refill sink or large pan with cold water. Drop tomatoes in boiling water, a few at a time. Scald for 30 seconds. Remove to cold water. Lift tomatoes from cold water. Peel and core tomatoes.
2. Pack tomatoes tightly in jars. Add boiling water or juice if necessary to leave ½ inch head space. Add salt (½ teaspoon per pint, 1 teaspoon per quart), if desired.
3. Load canner. Heat on high until a steady stream of steam flows out of the canner for 5 minutes.
4. Process in steam canner for 35 minutes for pints, 45 minutes for quarts.

Total Time: 2 hours for 7 quarts (21 pounds)

Note: You can do all of the above in a boiling water bath canner; but your time will be slightly longer, because it will take longer for the water in the boiling water bath to return to boiling when loaded then it takes for the steam canner to be hot enough to process. Processing times are the same.

Winter Squash and Pumpkins

Squash and pumpkins for winter storage should not be picked until they are fully mature and have attained the full color common to their type. Skins should be too hard to be pierced with your thumbnail. Leave at least 1 inch of stem on the fruit when you pick it.

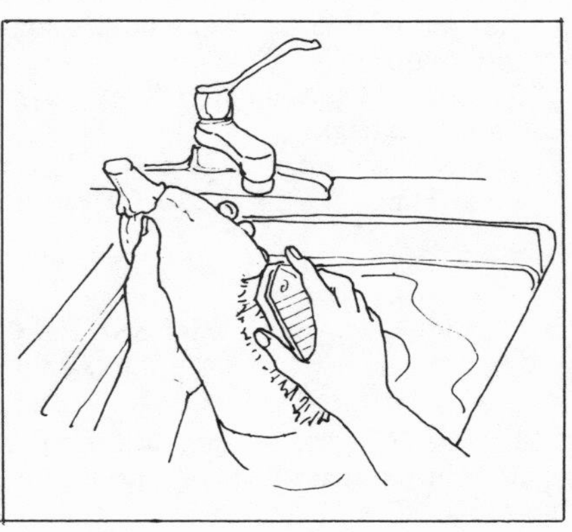

COOL STORAGE

Fastest, Best Finished Product

1. When harvesting squash and pumpkins, leave them in the sun or warm room (65–75 degrees F.) for 2 weeks to cure. This will make rinds harder. Acorn squash should be stored without curing.

2. Wash the vegetables with a solution of ½ cup chlorine bleach to a gallon of cold water. This will prevent bacteria growth during storage.

3. Store in a dry, well-ventilated area at 50–60 degrees F. Leave space between fruit to allow good air circulation.

Total Time: 10 minutes per fruit

Note: Squash and pumpkins should keep well into midwinter when any excess can be frozen or canned.

For best flavor and texture, reheat frozen squash in boilable bags for 20 minutes or in a double boiler until piping hot.

FREEZING SQUEEZO-STRAINED SQUASH IN BOILABLE BAGS

Fast Method, Excellent Product

Steam squash and pumpkins for freezing or canning when the woodstove is going in the winter.

1. Wash squash, cut in quarters or smaller. Steam until tender, 45–60 minutes.
2. Scoop out seeds. Scoop out pulp into Squeezo strainer and puree (or mash with a potato masher).
3. Pack in 1-pint boilable freezer bags. Add butter and seasoning, if desired. Press out air. Seal bags.
4. Cool. Wipe dry. Freeze.

Total Time: 1½–1¾ hours for 8 pints (12 pounds)

PRESSURE CANNING ONLY

When You Can't Freeze

Use an ice cream scoop to remove seeds and scoop pulp from rind.

1. Wash squash, cut in quarters or smaller. Preheat 2 quarts of water in canner.
2. Scoop out seeds. Scoop pulp into Squeezo strainer and puree (or mash with a potato masher).
3. Pack in jars. Leave ½ inch head space. Add salt (½ teaspoon per pint, 1 teaspoon per quart), if desired.
4. Load pressure canner. Add boiling water according to manufacturer's directions.
5. Exhaust canner according to manufacturer's directions, usually 7–10 minutes.
6. Process at 10 pounds: 65 minutes for pints, 85 minutes for quarts.
7. Allow canner to cool.

Total Time: 4½ hours for 7 pints (10 pounds)

Mixed Vegetables for Soups and Purees

At the end of the harvest season there is often an abundance of vegetables that have gone past their prime. These vegetables could be composted or fed to the pigs, but freezing or canning vegetables for sauces and soups are even better ways to make use of them.

FREEZING MIXED SOUP VEGETABLES

Fastest, Best Finished Product

1. Wash 16–20 pounds of vegetables. Peel if necessary. Slice, dice, julienne, or chop coarsely.

2. Mix the vegetables in a large pan. Add water to cover. Bring to a boil. Shut off heat.

3. Cool quickly for 30 minutes.

4. Stir well; pack in rigid containers. Leave 1 inch head space. Seal. Freeze.

Total Time: 2¾ hours for 7 quarts (16–20 pounds)

To freeze or can purees, chop the vegetables and cook in a minimum of water until mushy tender. Do not add seasoning. Puree in a Squeezo strainer, blender, food mill, or fine sieve. To freeze: cool quickly and pack in rigid containers; leave 1 inch of head space. To can: pack hot in jars, leave appropriate amount of head space, using individual vegetable instructions as a guide. Process for the length of time given for the longest processed vegetable plus 5 minutes to allow for the density of the puree.

PRESSURE CANNING ONLY

When You Can't Freeze

1. Wash the vegetables. Peel if necessary. Slice, dice, julienne, or coarsely chop.
2. Heat 2 quarts of water to boiling for canner and 2 quarts of water for filling jars.
3. Pack jars (layer for a pretty effect). Add boiling water to cover. Leave 1 inch head space. Add salt (½ teaspoon per pint, 1 teaspoon per quart), if desired.
4. Load canner and exhaust according to the manufacturer's directions, usually 7–10 minutes.
5. Process at 10 pounds: 60 minutes for pints, 90 minutes for quarts.
6. Allow canner to cool.

Total Time: 3¼ hours for 7 quarts (16–20 pounds)

CHAPTER 5

What Went Wrong? (And Other Commonly Asked Questions)

It's very discouraging to have anything go wrong when you are preserving food. Following directions carefully is the best way to avoid failures, but many factors can lead to a failure. The following commonly asked questions should help to solve—and prevent—many of your problems.

Q. Should all vegetables be blanched before freezing?

A. It has been thought up until now that all vegetables, except onions and peppers, should be blanched before freezing. However, we have discovered that some vegetables are best unblanched. These are beans, young broccoli, chopped peppers, and sliced summer squash. Corn in husks, whole tomatoes, and chopped onions also can be frozen unblanched.

Q. My frozen green beans are slimy and tough. How can I overcome that?

A. Green beans should not come in contact with water any more than necessary. Best results are obtained by freezing green beans that are unblanched or have been blanched in boilable bags for 6–8 minutes. They should be cooked by steaming or frying.

Q. When do you start counting time when you blanch vegetables?

A. When blanching in boiling water (immersion blanching) timing is started as soon as water returns to a boil. If it takes longer than 2 minutes for the water to boil, you have blanched too much at once. Reduce the amount of vegetables in the next batch. Steam blanch timing starts as soon as the cover is placed on the pan.

Q. What is the best way to freeze Sugar Snap peas?

A. Sugar Snap peas should be steam blanched for 3 minutes and tray frozen. Stir-fry Sugar Snaps for best cooked flavor.

Q. Sometimes my canning jars do not seal. Why?

A. Jars that come with mayonnaise, peanut butter, or commercially packed pickles may not fit the canning lids properly. Other reasons for unsealed jars are dented or rusty screw bands; nicked or cracked jars; handling lids with greasy fingers; not screwing bands on tightly enough; food, seeds, or herbs caught between the jar and the lid when sealing or when liquid is lost from the jars during the processing; or insufficient heat during processing.

Q. What causes jars to crack in the canner?

A. Jars will crack if they are overfilled. Using a metal instrument or table knife to expel air will weaken the bottom of jars causing them to fall off during processing.

Q. Must glass jars and lids be sterilized by boiling before canning?

A. No, not when boiling water bath or pressure canners are used. The containers as well as the food are sterilized during processing. But be sure jars and lids are clean, and prepared according to manufacturers directions before sealing jars.

Q. Why is liquid lost from jars during processing?

A. Loss of liquid may be due to packing jars too full, fluctuating pressure in a pressure canner, or lowering the pressure too suddenly. Failure to release air bubbles before sealing the jar will force out liquid when it begins to boil.

Q. Should lost liquid be replaced?

A. No, never open a jar and refill with liquid; this would let in bacteria and you would need to process again. Loss of liquid does not cause the food to spoil, though the food above the liquid may darken.

Q. How can I tell if my canned food is spoiled?

A. The signs of spoilage are bulging lids, broken seals, leaks, change in color, foaming, unusual softness or slipperiness, spurting liquid when the jar is open, mold, or bad smells. If any of these signs are noted, do not use the food. Discard it where animals and humans cannot find it.

Q. What makes canned food change color?

A. Darkening of foods at the tops of jars may be caused by oxidation due to air in the jars or by too little heating or processing to destroy enzymes. Overprocessing may cause discoloration of foods throughout the containers. Iron and copper from cooking utensils or from water in some localities may cause brown, black, and grey colors in some foods. When canned corn turns brown, the discoloring may be due to the variety of corn, to the stage of ripeness, to overprocessing, or to copper or iron pans.

Q. Is it safe to eat discolored foods?

A. The color changes noted above do not mean the food is unsafe to eat. However, spoilage may also cause color changes. Any canned food that has an unusual color should be examined carefully before use.

Q. Is it safe to use home canned food if the liquid is cloudy?

A. Cloudy liquid may be a sign of spoilage. But it may be caused by the minerals in hard water, or by starch from overripe vegetables. If liquid is cloudy, boil the food for 15 minutes before serving. Do not taste or use any food that foams during heating or has an off odor.

Q. What causes an accumulation of black material on the underside of some lids?

A. Natural compounds in some food cause deposits to form on the underside of lids. Unless the jar had sealed and then became unsealed, this deposit is harmless. Do not use a jar of food that has become unsealed.

Q. What causes a milky colored film on the outside of my jars?

A. Hard water leaves a mineral deposit on the outside of jars. To prevent this add ½ cup vinegar to each cannerful of water in hard water areas.

Q. Is it safe to can food without salt?

A. Yes, salt is used for flavor only and is not necessary for safe processing. Salt does help to retain the bright color of vegetables.

If that is a concern of yours, you can add a small amount of ascorbic acid to the jars. (I use half of a 50-mg. tablet.)

Q. Is it all right to use preservatives in home canning?

A. No. Some canning powders or other chemical preservatives may be harmful.

Q. Is processing time the same no matter what kind of stove you use?

A. As long as water in a boiling water bath canner or steam canner remains boiling throughout the processing time, or pressure remains stable throughout the processing time in a pressure canner, the processing times are the same.

Q. Is it safe to can foods in the oven?

A. No, oven canning is dangerous. Jars may explode. The temperature of the food in jars during oven-processing does not get high enough to insure destruction of spoilage bacteria in vegetables.

Q. Why is open-kettle canning not recommended for fruits and vegetables?

A. In open-kettle canning, food is cooked in an ordinary kettle, then packed into hot jars and sealed without processing. For vegetables, the temperatures obtained in open-kettle canning are not high enough to destroy all the spoilage organisms that may be in the food. Spoilage bacteria may get in when the food is transferred from kettle to jar.

Q. Is it safe to process sauces that contain low-acid foods in a boiling water bath canner or steam canner?

A. If the proportion of low-acid foods to acid foods is very low; for example, ½ cup each of onion and green pepper to a 7-quart batch of spaghetti sauce would be safe. Dry herbs and onion and garlic powder are preferable to fresh herbs when canning by these methods. Process these sauces for the same length of time as the acid food.

Q. Why do tomatoes sometimes float in the jar?

A. Tomatoes may float if they are overripe, packed too loosely, overprocessed, or if the temperature in a pressure canner becomes too high.

Q. Is it safe to can overripe tomatoes in a boiling water bath or steam canner?

A. Overripe tomatoes may be too low in acid to be safely canned by these methods. To be sure that the acid content is high enough (a pH rating below 4.5), purchase some nitrazine paper from a drugstore and dip a small piece into a cup of cooked tomatoes and test the acidity. Instructions come with the paper. If the pH is too high, that batch of tomatoes should be processed in a pressure canner at 10 pounds. Or you can add ¼ teaspoon of powdered citric acid per pint (½ teaspoon per quart) or 1 tablespoon white distilled vinegar per pint (8 tablespoons per quart) and process in a boiling water bath.

Q. Why do the canning lids pop off when I can tomato juice that was liquified in a blender?

A. Blenders incorporate a great deal of air into the vegetables they liquify. The juice should be heated to boiling and simmered 5 minutes to exhaust that air before canning. This does not lengthen the amount of time it takes to can tomato juice, since hot tomato juice is processed for only 10 minutes per pint and 15 minutes per quart. Juice and purees from Squeezo strainers should be handled the same way.

Q. May I use pressure canner for processing fruits and tomatoes?

A. Yes. If it is deep enough, it may be used as a water bath canner. Or you may use a pressure canner to process fruits and tomatoes at 0 to 1 pound pressure, without having the containers of food completely covered with water. Put water in the canner to the shoulders of the jars, fasten the cover. When live steam pours steadily from the open vent, start counting the time. Leave the vent open and process for the same time given for the boiling-water bath.

Q. My pressure canner does not have a pressure gauge. How can I tell if the correct pressure is being maintained?

A. As long as the control jiggles occasionally—as seldom as once every minute or two, that is your assurance that pressure is being maintained. The hissing sound and the slight escape of steam around the control, which are noticeable between jiggles, are additional assurances that proper pressure is being maintained. If the control jiggles too often, excessive loss of moisture will occur.

Q. What causes the safety fuse on a pressure canner to release?

A. It either pops out because the vent tube has become clogged with food or because of insufficient water in the canner.

Q. How do I know when my pressure canner needs a new gasket?

A. If steam persistently escapes around the rim of the cooker and pressure will not build up, a new gasket is needed. It should be replaced if it has become hard and slippery, or if it has stretched and will no longer fit easily into the cover.

Q. It seems to take excessively long for the pressure to go down in my pressure canner. Why?

A. As a rule it should only take between 45 and 60 minutes for the pressure to go down in a pressure canner. If your canner does not have a pressure gauge, you are most probably misjudging the test for steam pressure. Nudge the control, and if steam spurts out, the pressure is not down yet; if no steam spurts out, remove the control. The important thing is to *see* the steam, not hear it. You will *hear* a hissing sound whenever the control is nudged, but if you do not see steam it is safe to remove the cover.

Q. What causes beets to bleed?

A. Cutting the tops and roots off beets that are to be cooked in water before processing (1 inch of top and entire root should be left on), or piercing the beet to test for doneness too frequently. Beets baked in the oven do not bleed.

CHAPTER 6

Pickles, Relishes, Sauerkraut, and Sauces

Variety is the spice of our life, and pickles and relishes spice up many of my homecooked meals.

There are many books on pickles and relishes, but this one has been written for the busy person. I have timed each recipe so you know what you are getting into, and I have tried to find shortcuts.

One shortcut discovery I am very proud of is my Busy Person's Relish-Sauce-Chutney recipe. I found that I could make an Indian Relish, can half and save half. To the half I saved, I added peach preserves to make a great tasting Sweet and Sour Sauce. I canned half of that. To the half remaining, I added nuts and raisins and ended up with a marvelous chutney (so good I like to eat it right out of the jar!). Three different sauces out of 1 basic recipe—that's the kind of shortcut I hope you'll experiment with in your kitchen.

Pickles

Traditionally there have been 3 basic methods of making pickles: brining in salt, used most commonly for dill pickles and sauerkraut; fresh-pack, quick-processed pickles, which can turn vegetables into spicy sweet or sour pickles; and relishes, made by chopping or grinding a mixture of vegetables, adding seasonings, vinegar, and sweeteners, and cooking to the desired consistency. New to picklers are freezer pickles. These are the easiest to make, but if your freezer space is at a premium, you will find the other recipes in this section are quick and easy to make as well.

Basic Ingredients for Making Pickles

Vegetables and Fruit. These should be blemish free, firm, and fresh. Small, slender cucumbers and zucchini make the best whole pickles. Medium-size cucumbers or zucchini make good sliced pickles. Larger vegetables that have gone past their prime can be used to make relishes and sauces. Do not use waxed cucumbers; brine cannot penetrate the wax coating. Be sure to remove all blossom ends from the cucumbers; otherwise an enzyme in these blossoms may cause the pickles to soften during fermentation.

Salt. Use pure coarse salt, pickling salt, or kosher salt. Iodized salt will darken pickles and should not be used.

Vinegar. Use a high-quality vinegar of at least 4–6 percent acidity. (Don't use homemade vinegars.) Cider vinegar gives pickles a mellow, fruity taste and will produce a darker pickle. While distilled vinegar gives a sharper, more acid taste, it is best to use if you want a light-colored pickle. Never reduce the amount of vinegar called for in a recipe. Weakening the vinegar could enable bacteria to grow in the brine. If a less sour taste is desired, add additional sweetener.

Sweeteners. Use either white or brown sugar or honey. Brown sugar will produce a darker pickle. Honey usually will cloud the brine.

Spices. Use only fresh spices. Old spices may cause the pickles to taste musty. Whole spices are recommended over powdered. The whole spices should be loosely tied in a cheesecloth or small muslin bag and removed from the brine before packing the pickles in jars to prevent discoloration. However, many of the best cooks I know put the spices right into the brine and pack them with the pickles. The flavor of the spice is stronger, and the color is darker, but not unpleasantly so.

Grape Leaves. In the old days, many cooks used alum for a crisp pickle, but the USDA no longer considers alum safe for use in pickles. Grape leaves, added to the brine or stuffed into the bottom of a canning jar, are a good substitute. Use 1 grape leaf for each teaspoon of alum called for in a recipe.

Equipment

The equipment needed for making pickles is similar to that required for canning. Do not use copper, brass, galvanized, or iron pots or utensils. These metals will react with the acids and salt used in pickles, causing color and taste changes.

Here is the list of equipment you will need: sharp paring knives, scrub brush, large bowls, measuring cups and spoons, colander, fine sieves, cheesecloth, cutting board, crocks, heavy plate, clean jars, steam or boiling water bath canner, preserving kettle or large roaster, pan containing screw bands and lids in hot water, wide-mouth funnel, ladle, large wooden and slotted spoons, timer, hot pads, heavy bath towels, jar lifter, nonmetallic spatula for expelling bubbles from jars.

Before you start making pickles, be sure your work area and all equipment are spotlessly clean. If your jars have not been prewashed, wash them. Check the jars for nicks and cracks.

Processing

The USDA recommends that all pickles be processed. If you feel that processing alters the texture of your pickles, then you may keep unprocessed pickles in the refrigerator for a few weeks. We definitely do not recommend storing unprocessed pickles on a shelf.

You can process pickles in either a steam canner or a boiling water bath canner. A steam canner will produce a crisper pickle because the overall time the jar is in the canner is shorter.

Step-by-Step Pickle Making

1. Preheat the canner, jars, and lids while preparing the recipe.

To preheat a steam canner, fill the well with 2 quarts of water and turn the heat on high. Set the jars, bottom side up, on the steam canner rack and cover.

To preheat a boiling water bath canner, fill the boiling water bath canner with 4–4½ inches of hot tap water. Set the jars on a rack inside the canner, bottom sides up. Turn the heat on high.

Place the lids in a small pan of water and heat. Shut off the heat, and allow the jars and lids to remain in the hot water until they are needed.

2. Wash your vegetables thoroughly.

3. Prepare the vegetables according to the recipe directions—dice, chop, slice, and so on.

4. Add seasonings, sweeteners, and vinegar to the vegetables, according to the recipe directions.

5. Cook, if necessary.

6. Pack the vegetables firmly in the hot jars, making sure that the pickling brine fills the jars to the level indicated in the recipe.

7. Release any air bubbles in the jar by inserting the spatula and running it between the contents and the side of the jar.

8. Add more brine, if needed to maintain proper head space.

9. Wipe the rim of the jar with a clean, damp cloth to remove any food particles, seeds, or spices.

10. Adjust the lids as the manufacturer recommends.

11. Load the canner and bring the water in the canner to processing temperature.

In a steam canner, place the filled jars on the rack in the preheated steam canner. Set the cover dome in place. Heat the canner on high to allow steam in the dome to reach temperatures high enough to sterilize. When a steady stream of steam flows out of the vent holes for 5 minutes, start counting processing times.

In a boiling water bath, place the filled jars on a rack in a preheated canner. Make sure the water covers the jars by at least 2 inches. Cover the canner and bring the water to a boil.

12. Process for the length of time indicated in the individual recipe. Make sure the water is boiling throughout the processing time.

13. When the processing time is up, carefully remove the jars from the can-

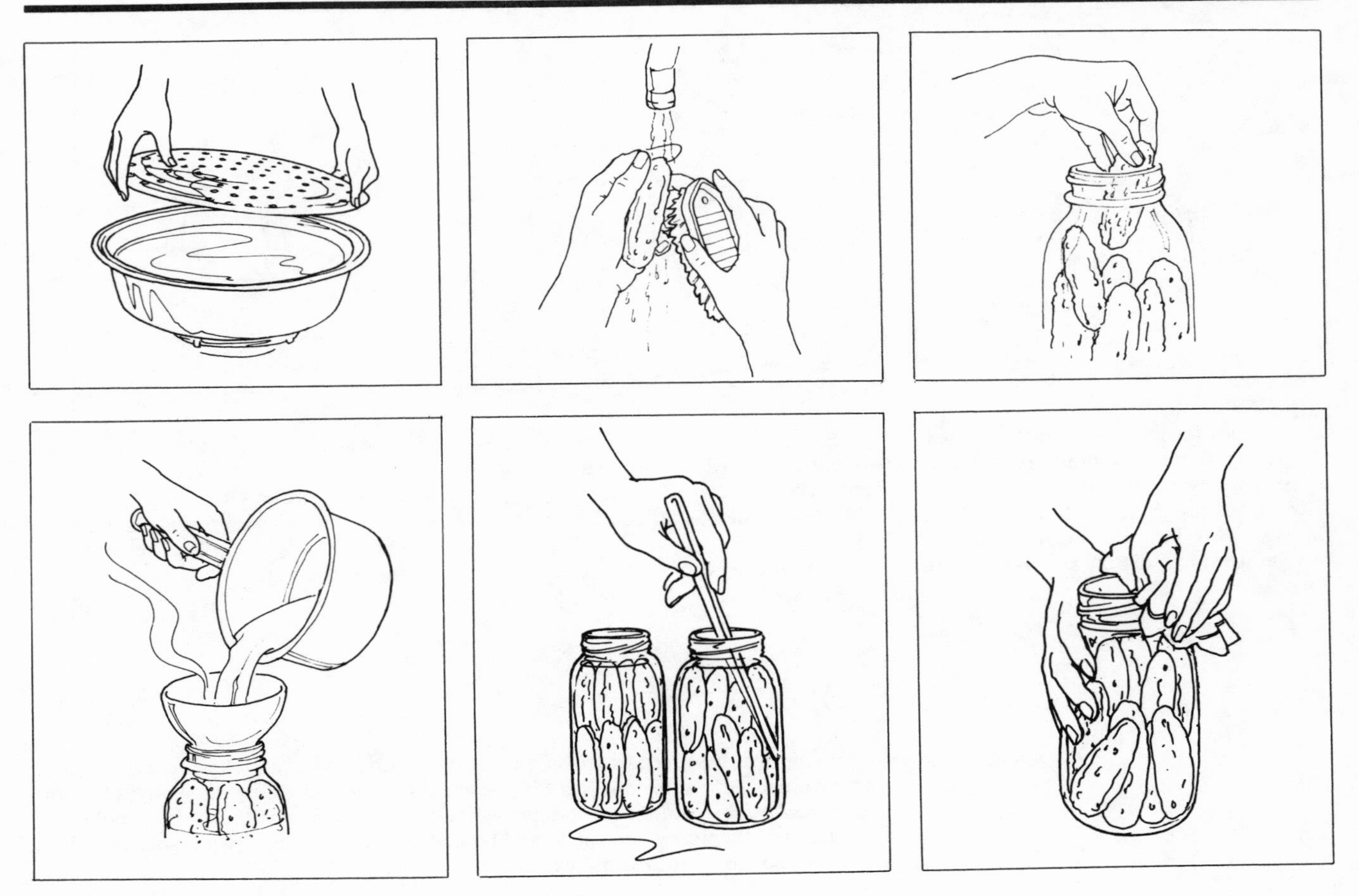

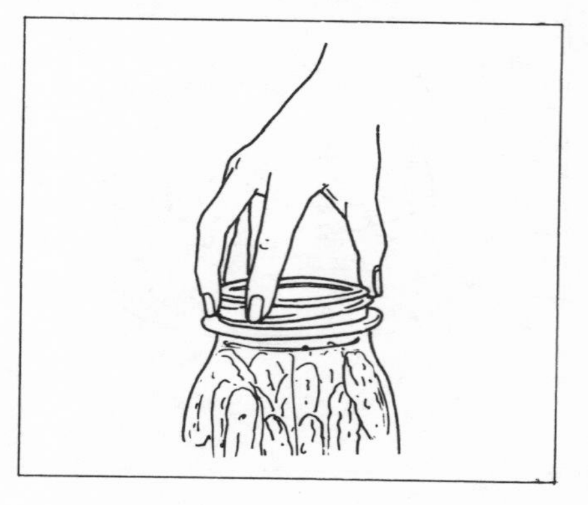

ner, using a jar lifter. Complete the seal according to the manufacturer's directions, if necessary.

14. Cool the jars for 24 hours. Check the seal. Any jars not sealed should be refrigerated immediately and used within 2 weeks.

15. Remove screw bands, wipe sealed jars, label, date, and store in a cool, dry, dark place.

Pickles should be chilled before serving.

When opening each new jar, check for signs of spoilage: bulging lids, leakage, mold, bad odor, very soft or mushy pickles. If there is the slightest doubt in your mind, *do not taste*. Dispose of the pickles where no person or animal can be harmed by them. Wash and sterilize the jar before storing or reusing.

Recipes

My timings for these recipes are based on using a food processor when necessary and a steam canner. Add additional time if you are chopping and slicing manually, or if you are using a boiling water bath canner. My times for each recipe include *actual work time*, and not the time when you would be free to be elsewhere, as when the vegetables are brining or the jars are cooling.

Prewash your jars and make up the pickling brine ahead. Heat the brine and can the dills as they come along, 2 or 3 quarts at a time.

SHORT-BRINE KOSHER DILL PICKLES

7 quarts
Approximately 1 hour

6 tablespoons pickling salt
4 tablespoons sugar
6 cups cider vinegar
6 cups water
7 large grape leaves
2 tablespoons mixed pickling spices
16–17 pounds pickling cucumbers (3–5 inches long)
21 large dill heads or 7 tablespoons dill seed
14 cloves of garlic
7 whole cayenne peppers (dried)

1. Preheat hot tap water in canner. Prepare lids. In a saucepan, mix the salt, sugar, vinegar, water, grape leaves. Add the whole pickling spices tied in a spice bag. Heat to boiling. Meanwhile, scrub the cucumbers.
2. Place 3 dill heads (or 1 tablespoon dill seed), 2 cloves garlic, and 1 cayenne pepper in each jar. Pack the cucumbers in the jars.
3. Fill the jars with hot brine to cover the cucumbers. Place 1 grape leaf in each jar. Leave ½ inch head space.
4. Process for 20 minutes in a preheated steam or boiling water bath canner. Start timing as soon as pickles go into canner.
5. Cool sealed jars. Check seals. Remove screw bands. Label. Store.

If you do not have enough small pickles for a full batch, pick the cucumbers daily as they come along and add to the brine. After each jar is filled, let it stand 7 days before processing.

SOUR PICKLES

7–8 pints
40–50 minutes (plus 7 days for brining)

Larger, slender cucumbers can be used for this recipe, but they should be cut in small chunks or spears.

2 quarts cider vinegar
½ cup dry mustard
½ cup sugar
½ cup pickling salt
60–80 scrubbed tiny cucumbers (1½–2½ inches long)

1. Combine the vinegar, mustard, sugar, salt. Pour into a clean gallon jar or container.
2. Add the cucumbers. Let stand for 7 days in a cool place.
3. Preheat hot tap water in the canner, prepare the lids. Meanwhile, drain the pickles and save the brine. Pack the pickles in clean jars. Fill the jars with the saved brine to cover the pickles. Leave ½ inch head space.
4. Process for 15 minutes in the preheated steam or boiling water bath canner. Start counting time as soon as pickles go into canner.
5. Cool sealed jars. Check seals, remove screw bands. Label. Store.

SWEET CHUNK PICKLES

4 pints
1½–2¼ hours (plus 10 days for brining)

8 cucumbers, 5 inches long
½ cup pickling salt (approximately)
1 quart cold water (approximately)
4 cups white vinegar
4 cups water
2 large grape leaves
4 cups sugar
1 tablespoon mixed pickling spices

1. Scrub the cucumbers. Cover with cold salt brine made of ½ cup salt for each quart of cold water used.
2. Let the cucumbers stand for 3 days.
3. Drain. Cover with cold water. Let stand for 24 hours.
4. Drain. Cover with cold water.
5. Let stand for 24 hours.
6. Drain. Cover with cold water.
7. Let stand for 24 hours.
8. Combine 2 cups vinegar, 4 cups water, and grape leaves. Heat to boiling. Meanwhile, cut the cucumbers into ½-inch chunks. Pour the hot brine over the cucumbers.
9. Let stand for 2 days.
10. Drain. Discard brine and grape leaves. Combine sugar, 2 cups vinegar, and spices. Heat to boiling. Pour over chunks.
11. Let stand for 24 hours.
12. Pour off brine and heat to boiling. Pour over chunks.
13. Let stand for 24 hours.
14. Fill canner with hot tap water. Preheat water and jars in canner. Prepare lids. Meanwhile reheat brine to boiling.
15. Pack the jars with pickles. Fill with hot brine. Leave ½ inch head space.
16. Place jars in preheated canner. Turn heat on high. Allow steam to flow out of vent holes of the steam canner in a steady stream for 5 minutes. Or bring the water to boil in a boiling water bath canner.
17. Process for 10 minutes.
18. Cool jars. Check seals, remove screw bands. Label. Store.

SUNSHINE PICKLES

7 pints
1½–1¾ hours

3 cups sugar
3 cups cider vinegar
1 tablespoon pickling salt
1 tablespoon celery seed
3 large grape leaves
5 quarts peeled and seeded ripe cucumbers or zucchinis (10 firm cucumbers or zucchinis, 6–8 inches long)
4 large onions, sliced
1 tablespoon turmeric

1. Combine the sugar, vinegar, salt, celery seed, and grape leaves in a preserving kettle. Heat on low while preparing cucumbers.
2. Peel and seed cucumbers; cut into "tongues" (2½–3-inch slices). Peel and slice the onions.
3. Place the vegetables in the hot brine. Bring to a boil, simmer until translucent. Meanwhile, fill the canner with hot tap water. Preheat jars and water in canner and prepare lids.
4. Remove the grape leaves; add the turmeric. Stir well. Pack the pickles in jars. Leave ½ inch head space. (If pickling zucchinis, add ⅓ grape leaf to each jar.)
5. Place the filled jars in preheated canner. Turn heat on high. Allow steam to flow out of vent holes of the steam canner for 5 minutes. Or, bring the water to a boil in a boiling water bath canner.
6. Process for 10 minutes.
7. Cool jars. Check seal, remove screw bands. Label. Store.

FREEZER PICKLES

4 pints
1 hour (plus 15 hours standing and chilling)

2 quarts sliced cucumbers (approximately 30 cucumbers, 5 inches long)
2 medium onions
2 tablespoons pickling salt
1½ cups sugar
1 cup cider vinegar
1 teaspoon celery seed

1. Scrub and slice the cucumbers. Peel and thinly slice the onions. Sprinkle with salt. Mix well.
2. Cover bowl and let stand for 3 hours.
3. Rinse the vegetables with cold tap water, and drain thoroughly. Meanwhile, mix sugar, vinegar, celery seed.
4. Pour the brine over the vegetables. Mix, cover. Refrigerate overnight.
5. Pack the pickles in straight-sided containers. Cover with brine. Leave 1 inch head space. Seal. Freeze. Defrost in the refrigerator for 8 hours before serving.

FREEZER COLESLAW

Approximately 4 pints
1¾ hours (plus standing and cooling time)

1 medium-size cabbage
1 carrot
1 green pepper
2 tablespoons pickling salt
1 cup cider vinegar
1¼ cups sugar
¼ cup water
1 teaspoon celery seed

1. Shred the cabbage, carrot, green pepper. Sprinkle with salt; mix well. Cover. Let stand for 1 hour.
2. Meanwhile, mix the vinegar, sugar, water, celery seed. Bring to a boil. Boil 1 minute.
3. Rinse the vegetables with cold tap water and drain. Squeeze out as much water as possible.
4. Pour the brine over the cabbage mixture. Stir well. Cool.
5. Pack in straight-sided containers. Leave 1 inch head space. Seal. Freeze. Defrost and store in refrigerator before serving.

PICKLED BEETS

7 pints
2¼–2¾ hours

10–12 pounds beets (without tops)
1 quart cider vinegar
⅔ cup sugar
1 cup water
2 tablespoons salt

1. Cut the tops and roots off flush with the beet. Scrub thoroughly.
2. Place the beets on a rack in a large roaster. Cover and bake at 400 degrees F. until tender, about 1 hour for medium-size beets. Meanwhile, preheat hot tap water and jars in canner. Prepare lids.
3. In a saucepan, mix the vinegar, sugar, water, salt. Heat to boiling.
4. When the beets are tender, fill the roaster with cold water. Slip the skins off the beets.
5. Pack the beets in hot jars, whole or cut. Add brine to cover. Leave ½ inch head space.
6. Put filled jars in the preheated canner. Turn heat on high. Allow steam to flow out of the vent holes of the steam canner in a steady stream for 5 minutes. Or, bring the water to a boil in a boiling water bath.
7. Process for 10 minutes.
8. Cool jars. Check seals, remove screw bands. Label. Store.

GERMAN PICKLE RELISH

Approximately 7 pints
2–2⅓ hours

2 quarts finely chopped or ground cabbage
1 quart ground green tomatoes
2 sweet red peppers
1 quart sliced onions
1 tablespoon celery seed
1 tablespoon pickling salt
¼ teaspoon pepper
12 whole cloves
½ cup mustard seed
4½ cups sugar
1 quart cider vinegar
1 teaspoon turmeric

1. Finely chop or grind the cabbage, green tomatoes, peppers, onions.
2. Combine the vegetables with the remaining ingredients, *except* turmeric, in a large saucepan. Heat to boiling. Simmer until cabbage is well done (about 1 hour). Meanwhile, preheat hot tap water and jars in canner. Prepare lids.
3. Remove relish from heat when done. Stir in turmeric.
4. Pack the relish in hot jars. Leave 1 inch head space.
5. Put filled jars in preheated canner. Turn heat on high. Allow steam to flow out of vent holes of the steam canner in a steady stream for 5 minutes. Or, bring the water to boil in a boiling water bath canner.
6. Process for 5 minutes.
7. Cool jars. Check seals, remove screw bands. Label. Store.

BUSY PERSON'S RELISH-SAUCE-CHUTNEY

14 pints and 7 half-pints
4–5 hours

This recipe makes 3 great tasting sauces: Indian Relish, a spicy sauce to eat with meats (makes a great Thousand Island Salad Dressing); Sweet and Sour Sauce, to use with meat, poultry, and seafood (especially good with Chinese or Polynesian dishes); and Chutney, fantastic with everything. It's very easy to make, but takes a little longer than other recipes. This is a good project for a free day.

Relish:

4 quarts tomato puree (approximately 24 large tomatoes)
24 large apples
7 cups chopped onions (approximately 18 large onions)
2 quarts cider vinegar
6 cups sugar
⅔ cup salt
2 teaspoons ground cloves
2 teaspoons red pepper
2 teaspoons cinnamon
2 teaspoons dry mustard

Sweet and Sour Additions:

7 cups peach preserves
2 teaspoons garlic powder
1 teaspoon Tabasco sauce

1. Puree tomatoes in a Squeezo strainer. Peel, core, and coarsely chop apples. Peel and chop onions. Mix in large roaster or preserving kettle with vinegar, sugar, salt, cloves, red pepper, cinnamon, dry mustard.
2. Bring to a boil and simmer until thick, about 2 hours. Meanwhile, preheat hot tap water and jars in canner. Prepare lids.
3. Ladle hot sauce into 7 hot pint-size jars. Leave ½-inch head space.
4. Put filled jars in the preheated canner. Turn heat on high. Allow steam to flow out of the vent holes of the steam canner for 5 minutes. Or, bring the water to a boil in a boiling water bath.
5. Process for 5 minutes. While the jars are processing, add the peach preserves to the remaining sauce and reheat to boiling.
6. Fill 7 pint-size jars, leaving ½ inch head space.
7. Put filled jars in the preheated canner. Turn heat on high. Allow steam to flow out of the vent holes of the steam canner for 5 minutes. Or, bring the water to a boil in a boiling water bath.
8. Process the jars for 5 minutes. While the jars are processing, add the raisins and nuts to the remaining sauce, and reheat to boiling.
9. Pack the remaining hot sauce into 7 ½-pint jars. Leave ½ inch head space.

Chutney Additions:

1 cup raisins
1 cup chopped walnuts

10. Put filled jars in the preheated canner. Turn heat on high. Allow steam to flow out of the vent holes of the steam canner for 5 minutes. Or, bring the water to a boil in a boiling water bath. Process for 5 minutes.
11. Cool jars. Check seals, remove screw bands. Label. Store.

ZUCCHINI RELISH

Approximately 7 pints
2¼–2½ hours (plus overnight brining)

10 cups finely chopped zucchini
4 cups finely chopped onions
1 green pepper, chopped finely
1 red pepper (sweet), chopped finely
5 tablespoons pickling salt
2½ cups white vinegar
1 large cayenne pepper with seeds
1 tablespoon nutmeg
1 tablespoon dry mustard
1 tablespoon turmeric
1 tablespoon cornstarch
½ teaspoon pepper
2 teaspoons celery salt

1. Chop the vegetables and sprinkle salt over them. Mix well. Let stand overnight.
2. Drain the vegetables. Rinse thoroughly with cold tap water. Drain again.
3. Place the vegetables in a large kettle with the remaining ingredients. (Puree the cayenne pepper in blender with a little of the vinegar for better flavor.) Bring to a boil. Simmer for 30–45 minutes until thick. Meanwhile preheat hot tap water and jars in canner. Prepare lids.
4. Pack jars. Leave ½ inch head space.
5. Put filled jars in preheated canner. Turn the heat on high. Allow steam to flow out of the vent holes of the steam canner for 5 minutes. Or, bring the water to a boil in a boiling water bath.
6. Process for 5 minutes.
7. Cool jars. Check seals, remove screw bands. Label. Store.

SAUERKRAUT

7 pints
2¼–2⅔ hours (plus 4 weeks for brining)

5 pounds cabbage
3 tablespoons pickling salt
12 juniper berries
½ cup Chablis wine

1. Shred the cabbage. Layer the cabbage, salt, and 3 juniper berries per layer in a large crock or bowl that holds at least 1 gallon. Tap every other layer with a potato masher to get rid of trapped air bubbles. Cover with a clean cloth and weigh down with a heavy plate. Place this container in another pan to collect fermenting juices that overflow. Place in an area that remains between 65 and 75 degrees F.
2. By the next day, brine will form and cover the cabbage. By the second day, scum will start to form. On the second day, pour the wine over all. Rinse the plate and cloth and replace each day.
3. Skim any scum that has formed after 2 weeks.
4. Skim again after 4 weeks.

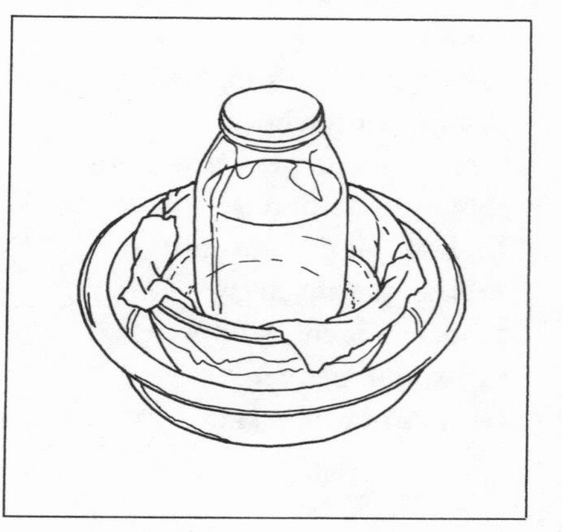

5. To can (hot pack only), fill the canner with hot tap water and preheat water and jars in canner. Prepare lids.

While the canner and jars are preheating, heat the sauerkraut to simmer. *Do not boil*. Pack in hot jars. Leave ½ inch head space. If you run short of juice, mix a brine of 1½ tablespoons salt to 1 quart boiling water. (Divide this mixture between the jars; do not use the new brine to fill just 1 jar.)

6. Put filled jars in the preheated canner. Turn heat on high. Allow steam to flow out of the vent holes of the steam canner for 5 minutes. Or, bring the water to a boil in a boiling water bath.

7. Process: 15 minutes for pints, 20 minutes for quarts.

8. Cool jars. Check seals, remove screw bands. Label. Store.

Sauerkraut can be stored in a crock, and not canned, if the area in which it is stored remains at a constant 38 degrees F. When removing a portion of sauerkraut from the crock, make sure that remaining sauerkraut is covered with brine. Mix more brine, 1½ tablespoons of pickling salt to 1 quart water, if necessary. Always use a glass or china cup to remove a portion of sauerkraut.

How can you tell if sauerkraut is spoiled? Spoiled sauerkraut has an off odor and changes in color. Soft kraut is caused by insufficient salt, uneven distribution of salt, temperature too high during fermentation, or not expelling air bubbles when packing. Pink kraut indicates a yeast growth on the surface, too much or improperly distributed salt, or not keeping the kraut weighted down during fermentation. Rotted kraut is caused by an insufficient covering of brine to exclude air. Dark kraut is caused by cabbage that was washed or trimmed improperly; insufficient covering of brine to exclude air; poor distribution of salt; high fermentation, processing, or storage temperatures; or kraut that has been in storage too long.

If your family does not care for the consistency of homemade catsup, mix the finished product with commercially prepared catsup in a 2 to 1 ratio, and they will never know the difference.

Oven Method of Cooking Sauces

Spaghetti sauce and catsup can be reduced on top of the stove by cooking for 3 hours, but you must watch the pot constantly to keep it from boiling over or sticking. It must be stirred often. Sauces cooked in the oven take longer to reduce; but they do not need to be stirred. They will not boil over, nor will they stick to the pot. While the sauce is in the oven, you are free to do other things. If you would like to hurry the process toward the end, you can quickly reduce the sauce by cooking it for a very short period of time on top of the stove.

JUST LIKE STORE-BOUGHT CATSUP

6 quarts
1–1½ hours (plus 10 hours cooking time)

8 quarts tomato puree (approximately 24 pounds tomatoes)
12 ounces thick tomato paste
3 green peppers
5 large onions
2 cups cider vinegar
1 cup light corn syrup
1 cup sugar
1 teaspoon pepper
2 tablespoons salt
2 teaspoons allspice

1. Puree the tomatoes with a Squeezo strainer. Liquify the tomato paste, peppers, and onions in a blender with some of the vinegar. Mix the purees with remaining ingredients in a large roaster. Stir well. Bring to a boil on top of the stove.
2. Cook, uncovered, in a 200-degree F. oven for 10 hours. Do not stir.
3. One hour before cooking time is up, fill canner with hot tap water and preheat water and jars in canner. Prepare lids.
4. When cooking time is up, ladle the hot catsup into the hot jars. Leave ½ inch head space.
5. Put the filled jars in the preheated canner. Turn heat on high. Allow steam to flow out of the vent holes of the steam canner for 5 minutes. Or, bring the water to a boil in a boiling water bath.
6. Process: 10 minutes for pints, 15 minutes for quarts.
7. Cool jars. Check seals, remove screw bands. Label. Store.

Should you decide to cook these sauces on the top of the stove, place a small, clean window screen over your roaster to prevent spatters. This will save clean-up time later.

JAN'S SPICY SPAGHETTI SAUCE

6–7 quarts
1–1½ hours (plus 10 hours cooking time)

10 quarts tomato puree (approximately 30 pounds tomatoes)
4 large onions
⅓ cup dried sweet basil
2 tablespoons dried oregano
1 teaspoon dried thyme
1 teaspoon pepper
¾ cup honey or sugar (optional)
1 teaspoon dried marjoram
5 bay leaves
1 tablespoon crushed red pepper
1 tablespoon garlic powder
2 tablespoons salt
1½ tablespoons dried parsley

1. Puree tomatoes in a Squeezo strainer. Chop onion finely. Put the vegetables with remaining ingredients in a large roaster. Stir well. Bring to a boil on top of the stove.
2. Cook, uncovered, in a 200-degree F. oven for 10 hours. Do not stir.
3. One hour before cooking time is up, fill canner with hot tap water and preheat water and jars in canner. Prepare lids.
4. When cooking time is up, ladle the hot sauce into hot jars. Leave ½ inch head space.
5. Put the filled jars in the preheated canner. Turn the heat on high. Allow steam to flow out of the vent holes of the steam canner for 5 minutes. Or, bring the water to a boil in a boiling water bath.
6. Process: 10 minutes for pints, 15 minutes for quarts.
7. Cool jars, check seals, label, store.

11 Commonly Asked Pickling Questions

Q. Must I process pickles and relishes?

A. Yes, the USDA now recommends processing all pickles and relishes to ensure destruction of harmful bacteria.

Q. Must I sterilize jars when I process pickles and relishes?

A. When pickles and relishes are processed, it is not necessary to sterilize jars.

Q. What causes soft or slippery pickles?

A. Not using freshly picked cucumbers, too little salt in the brine, acidity of vinegar is less than 5 percent, reducing the acid strength of the vinegar by adding more water than the recipe called for, not removing scum regularly, vegetables not covered with brine, hard water, not removing blossom ends from cucumbers, processing too long, or storing pickles where it is too warm.

Q. What causes hollow pickles?

A. Growing conditions (long dry spells followed by heavy rains), cucumbers stored too long before pickling, brine not strong enough, or high fermentation temperatures.

Q. What causes shriveled pickles?

A. Cucumbers stored too long before pickling, pickling solution too sweet or vinegar too strong, not enough salt in the brine, cooking too long, or processing too long.

Q. What causes dark pickles?

A. Using too much spice, not removing whole spices from brine before packing pickles, minerals in the water (hard water), overcooking, cooking in an iron kettle, using iodized salt, or low nitrogen content of cucumbers.

Q. Why do my pickles become dull or faded?

A. Poor quality cucumbers, sunburned or overmature cucumbers, or poor growing conditions.

Q. What is the white sediment that collects in the bottom of my jars of pickles?

A. Bacteria caused by fermentation, fluctuations in storage temperatures, or not using pure canning salt. It is usually not harmful unless the pickles are spoiled. If the pickles are spoiled, the jars frequently will spurt liquid when opened.

Q. What causes pickles to spoil?

A. Not following directions carefully for packing and processing, using ingredients that are too old, or weakening the vinegar solution by adding more water than the recipe called for.

Q. Now that the USDA no longer approves of the use of alum, how do we keep our pickles crisp?

A. Use grape leaves (cultivated or wild) according to your recipe directions, or, to adapt your recipes, substitute 1 large grape leaf for each teaspoon alum called for in the recipe. Except with dill pickles, remove the grape leaves before packing. Do not overprocess pickles.

Q. When I add whole garlic to my dills the garlic turns blue. Why?

A. Hard water is usually the cause.

CHAPTER 7

Quick Harvest Meals

I have 2 pieces of advice for making harvest mealtimes easier: incorporate the vegetables you are processing into your meals and have plenty of basic foods in the freezer, ready to be converted into instant meals.

Instant Meals

As winter slips into early spring, prepare extra meals to tuck away in the freezer for harvest dinners without hassle. You can freeze ahead dishes from recipes in this chapter, as well as individually frozen hamburg patties, beef and chicken stocks and stew bases, crepes, pie crusts for quick meat pies or vegetable quiches, and favorite casseroles.

To save time in meal preparation, incorporate the vegetables you are working with into your meal plan for that day or the following day. For example, if you are canning or freezing spaghetti sauce, plan a spaghetti dinner. Remove from the freezer the number of servings of Freezer Meatballs (p. 109) or frozen cooked ground beef that you need, or brown some fresh beef. While the meat is heating in a pan of spaghetti sauce, prepare the pasta, and you have an almost instant dinner.

Diced Meats: The Basis of 13 Quick Meals

A supply of diced, cooked poultry, beef, pork, and ham frozen with their cooking broths can be the basis of a variety of different meals. Tender cuts can be roasted, cooled, diced, and packed with broth made from pan juices; less tender cuts should be simmered.

Roasts of beef, pork, ham, veal, chicken, and turkeys up to 12 pounds can all be prepared by slow cooking in the oven—while you are busy elsewhere. To do so, prepare your roast as usual; do not stuff roasting chickens or turkeys. Early in the morning, roast the meat at the normal roasting temperature for 30 minutes. Reduce the oven temperature to 185 degrees F. (150 degrees F. for medium-rare roast beef and 200 degrees F. for a 10–12 pound turkey) and roast, uncovered, for 8–10 hours. The meat will be perfect for supper, and if you are a little late, it won't overcook.

Pot roast can be prepared the same way, adding vegetables if desired. Roast the meat covered for ½ hour at 375 degrees F., reduce the temperature to 185 degrees F., and roast 8–10 hours. Add seasonal vegetables, and you have a complete meal and enough leftovers to freeze for some instant meals.

Packages containing 2 cups of meat covered with broth will give you the greatest versatility. Here are just a few ideas for using these meats. I'm sure you will come up with more of your own.

- Combine with soup stock and seasonal vegetables for a hearty stew
- Stir-fry with vegetables, add walnuts for variety
- Make into hash
- Cream chicken with new peas and carrots, and serve over hot biscuits
- Make sandwiches, salads, quiches, and casseroles
- Mix with homemade Indian relish, serve on hamburger rolls

- Mix with homemade chutney, serve over rice or noodles
- Add to cooked vegetables and stock, thicken with flour, and top with a crust or biscuits for a quick meat pie
- Make chow mein
- Mix with vegetables and eggs for a quick stove-top frittata
- Mix with rice and vegetables for stir-fried rice

Freeze-Ahead Recipes

The following recipes are for foods that can be prepared in advance to save meal preparation time during the busy harvest season. Serve part the day you make it and freeze the rest. Most are basic recipes that later can be used to create a variety of nourishing and delicious dishes using the sauces and vegetables you are preserving.

To save preparation time and energy, defrost frozen foods in the refrigerator overnight unless otherwise specified.

Preparation times given for all recipes in this chapter are based on the use of a food processor, blender, and electric mixer when preparing foods. If you do not use these appliances, be sure to add additional time. Also, when I added up the total work time, I did not include cooking, baking, and chilling times when I was free to perform other tasks.

3 IN 1 CHICKEN RECIPE

24 servings
2¼ hours (plus time for the chicken to cool and freeze)

12 pounds chicken fryer parts
1 tablespoon salt
1 large onion, cut in chunks
3 cups unbleached flour
3 cups crushed cornflakes
1¼ teaspoons baking powder
5 teaspoons salt
1 tablespoon garlic powder
1½ teaspoons dried tarragon leaves
1½ teaspoons dried chives
1½ teaspoons parsley flakes
3 eggs
¾ cup cold milk
2 teaspoons onion powder
peanut oil for frying

1. Place the chicken in a large kettle with the salt and onion and water to cover. Bring to a boil, and simmer for 20 minutes.
2. Turn off the heat, and cool the chicken in the cooking broth (about 3 hours).
3. Drain well and save the stock for soup.
4. While the chicken is cooling, mix the flour, cornflake crumbs, baking powder, and seasonings. Set aside. Blend the eggs, milk, and onion powder. Set aside.
5. In a large, heavy skillet or electric frying pan, heat 2 inches of oil to 380 degrees F. Dip the chicken pieces in the egg mixture and then in the flour mixture. Fry until tender and golden brown on all sides. Remove the chicken from the frying pan, and drain on paper towels. It takes about 1½ hours to fry 12 pounds of chicken.
6. Cool completely (about 1 hour).
7. Tray freeze for 12 hours.
8. When the chicken is frozen solid, package in large freezer bags to be used as needed.

FRIED CHICKEN

Place frozen chicken parts on a well-drained baking sheet, and heat uncovered in a preheated, 400-degree F. oven for 30–45 minutes. (Time depends on thickness of the chicken parts.)

POLYNESIAN CHICKEN

While 8 pieces of chicken are heating in the oven as directed above, sauté 1–2 cups of vegetables (carrots, broccoli, onions, peppers) in 1 tablespoon of cooking oil until tender. Add 2 cups of homemade chutney (p. 94) and heat through. Remove the cooked chicken from the oven, and top with the vegetable-chutney sauce. Return to the oven and bake an additional 10–12 minutes. Serves 4 with rice.

BARBECUED CHICKEN

While 8 pieces of chicken are baking as in the directions given with the Fried Chicken, sauté 1 medium diced onion in 1 tablespoon oil until tender. Add 1 cup homemade catsup, 1 teaspoon honey, 1 tablespoon prepared mustard, 1 teaspoon salt, 1 ½ tablespoons cider vinegar, ½ teaspoon Worcestershire sauce, and ¼ teaspoon hot pepper sauce. Simmer for 20 minutes. When the chicken is heated through, top with the barbecue sauce and return to the hot oven for an additional 10–12 minutes. Serves 4.

CHILI

25 servings
3¼ hours

4 pounds lean ground beef
2 tablespoons salad oil
3 cups diced onions
3 cups diced green pepper
8 cups cooked, drained kidney beans
2 quarts canned tomatoes
1½ cups tomato paste
1 cup water
2 tablespoons chili powder
½ teaspoon cayenne powder
½ teaspoon paprika
3 large bay leaves, crushed finely

1. Brown the beef in oil until it loses its red color. Add the remaining ingredients.
2. Cook uncovered until thick, about 2 hours. Stir occasionally.
3. Cool for 1 hour, stirring often.
4. Package in meal-size portions in straight-sided containers. Leave a 1-inch head space. Label. Freeze.

Defrost the chili overnight in the refrigerator or run cold water over the container to remove the frozen chili, and reheat in the top of a double boiler.

Serve with diced raw onion and grated Parmesan cheese, or serve in taco shells topped with diced onion, lettuce, and shredded cheddar cheese.

MEAT LOAF

2 large or 4 small loaves
15 minutes (plus baking and cooling time)

3 pounds ground beef
1 pound ground pork
4 eggs
2 large onions, cut in chunks
6 ounces of bread (about 6 slices) torn in small pieces
1½ cups milk
¾ teaspoon dry mustard
1 teaspoon garlic powder (optional)
1½ teaspoons thyme
4 teaspoons salt
½ teaspoon pepper

1. Preheat the oven to 325 degrees F. Place the beef and pork in a large bowl. Place the rest of the ingredients in a blender or food processor. Blend until smooth. Pour this over the meat, and mix thoroughly. Divide into two 9″ × 5″ × 3″ loaf pans, or four 9″ × 4″ × 2½″ pans.
2. Bake for 1½ hours (in large pans) or 1 hour (in small pans).
3. Remove the loaves from the oven. Drain excess fat. Cool completely.
4. Remove the loaves from the baking pans and wrap for the freezer. Label. Freeze.

Defrost overnight in the refrigerator and serve cold, surrounded by fresh vegetables or in sandwiches. Or reheat in a 350-degree F. oven for 20–30 minutes.

You can top the meat loaf with spaghetti sauce, Sweet and Sour Sauce, or hot mashed potatoes covered with a slice of your favorite cheese. Slip the meat back into the oven until the cheese melts or the sauces are heated through.

QUICK AND EASY LASAGNA

12 large servings
1 ½ hours (plus time to make the sauce and freezing time)

4 cups spaghetti sauce (p. 99 or use your favorite recipe)
1 pound uncooked lasagna noodles
2 pounds cottage cheese
2 cups sour cream
1 pound mozzarella cheese, sliced thinly
1 cup grated Parmesan cheese
2 cups water

1. Preheat the oven to 350 degrees F.
2. Foil line 2 casserole dishes. Spread 1 cup of spaghetti sauce in the bottom of each dish. In a bowl, mix the cottage cheese with the sour cream. Layer the ingredients in the casserole dishes, beginning with the noodles, followed by the cottage cheese and sour cream mixtures, then the mozzarella, ½ cup sauce, and finally the Parmesan cheese. Repeat 1 more layer, dividing the ingredients evenly between the casserole dishes. Pour 1 cup water around the sides of each dish.
3. Cover tightly with foil and bake for 1 hour. If both casseroles are to be frozen, remove from oven after 1 hour. If a pan of lasagna is to be eaten, bake it uncovered an additional 15 minutes or until the noodles are tender. Let it stand 15 minutes before serving.
4. Cool for 2 hours.
5. Freeze.
6. Remove the casserole from the baking dish. Wrap with freezer wrap or foil. Return to the freezer.

Return to a baking dish and defrost in the refrigerator overnight. Bake covered in a preheated, 350-degree F. oven for 30–40 minutes. Let the lasagna stand 15 minutes before serving.

FREEZER MEATBALLS

Approximately 100 meatballs
1½ hours

5 pounds lean ground beef
3 eggs
1½ cups quick-cooking oatmeal
¾ cup grated Parmesan cheese
3 large cloves garlic
2 large onions, coarsely chopped
1½ teaspoons dried oregano
2 tablespoons dried basil
2 teaspoons crushed red pepper
1 tablespoon salt
¾ teaspoon black pepper

1. Place the beef in a large bowl. Put the eggs, oatmeal, cheese, garlic, onions, herbs, and spices in a blender or food processor. Puree until smooth. Add to the beef and mix thoroughly with your hands.
2. Preheat the oven to 400 degrees F.
3. Shape the meat mixture into 1-inch balls. Place on lightly greased cookie sheets.
4. Bake for 25 minutes at 400 degrees F.
5. Cool at room temperature for 30 minutes. Drain fat.
6. Tray freeze. When frozen, pack in large plastic freezer bags. Freeze.

For a quick 15-minute supper on the day you are making Jan's Spicy Spaghetti Sauce, ladle some sauce into a saucepan; add the desired number of frozen meatballs. Simmer until the meatballs are cooked through (about 15 minutes). Meanwhile, prepare some quick-cooking vermicelli (Mueller's has one that cooks in 5 minutes). Serve with fresh garden salad.

Another quick meal can be made by heating the meatballs in spaghetti sauce, and serving on hamburger rolls for hot meatball sandwiches.

BAKED BEANS

10–12 servings
45 minutes (plus overnight soaking and 5 hours baking)

2 pounds navy pea beans
4 quarts cold water
1½ cups brown sugar
2 teaspoons dry mustard
1 teaspoon salt
½ teaspoon pepper
½ teaspoon hot pepper sauce
¼ cup tomato paste
boiling water
12 small onions
12 slices fresh (uncured) bacon

1. Rinse the beans with cold water. Place in a 10-quart kettle. Add 4 quarts of cold water. Let stand overnight.
2. Cook the beans in the water they were soaked in. Cover the kettle; bring the beans to a boil. Then lower the heat and simmer until tender. Skim off the foam as it rises to the top.
3. Drain the beans. Preheat the oven to 325 degrees F.
4. Line two 2-quart baking dishes with foil. Divide the beans between the two baking dishes. Put half of the dry mustard, salt, pepper, hot pepper sauce, and tomato paste in each dish. Cover the beans with boiling water. Stir well. Top with onions; then fresh bacon slices.
5. Bake, uncovered, for 5 hours, adding water as necessary.
6. Cool at room temperature for 1½ hours.
7. Freeze for 12 hours.
8. Remove the frozen beans from the baking dishes. Overwrap with freezer wrap or foil. Label and return the packages to the freezer.

Defrost frozen beans in the refrigerator. Place the foil package in a baking dish. Bake, uncovered, in a 325 degree F. oven for 1 hour. Add a small amount of boiling water, if necessary. Or place the frozen casserole back into a casserole dish. Put in a 325-degree F. oven and bake covered for 1 hour, uncover, and bake another hour. Add boiling water if necessary.

Cold baked beans can be mashed, with raw onion to taste, for a sandwich filling. This is delicious served on rye bread with a spicy mustard.

Harvest-time Dishes

In the thick of preserving vegetables, you can save time by cooking a dinner based on the vegetable you are processing. Base the whole meal around the vegetable, or use a vegetable dish to supplement some of the casseroles and other goodies you have tucked away in the freezer for busy days.

Here are a few of my favorite vegetable recipes.

MAGIC CRUST BROCCOLI PIE

4–6 servings
1 hour

2 cups chopped cooked broccoli
1 cup chopped cooked ham
¾ cup shredded cheddar cheese
½ cup unbleached flour
⅛ teaspoon salt
¾ teaspoon double-acting baking powder
1 tablespoon shortening
2 eggs
¾ cup milk

1. Preheat the oven to 350 degrees F.
2. Layer the broccoli, ham, and cheese in a well-oiled 9-inch pie plate. Blend the remaining ingredients in a blender, food processor, or with an electric hand mixer. Pour over the cheese layer.
3. Bake, uncovered, for 40 minutes.
4. Let stand for 10 minutes before serving.

VEGETABLE PLATTER WITH PEANUT SAUCE

4 servings
1 hour

Vegetables:

4 large carrots (scrubbed and peeled, if desired)
4 medium potatoes, pared
½ medium head of cabbage, cut in wedges
2 cups whole green beans
1 medium-size bunch of broccoli, separated into slender stalks
1½ cups water

Sauce:

¾ cup peanut butter
4 tablespoons chopped peanuts
½ cup beef stock, bouillon, or water
½ cup coffee cream
1 tablespoon lemon juice
1 clove garlic, minced finely
¼ teaspoon crushed red pepper

1. Prepare the vegetables.
2. Bring the water to a boil in a large kettle. Add the carrots and potatoes. Reduce heat, and cook for 15 minutes.
3. Add the cabbage, beans, and broccoli. Continue cooking until the potatoes are tender and the other vegetables are tender crisp, about 15 minutes.
4. Drain the liquid, and keep the vegetables warm. Meanwhile, combine the sauce ingredients in a medium-size saucepan, and cook over medium heat until heated through.
5. Arrange the vegetables on a large platter and top with the sauce.

CHEESE STUFFED PEPPERS

4 servings
45 minutes

4 large green peppers, seed pods removed
1 large tomato, scalded, peeled, and cubed
2 teaspoons fresh basil, finely chopped, or ½ teaspoon dried
¼ teaspoon salt
dash pepper or cayenne
½ pound sharp cheddar cheese, cut in ¼-inch cubes
½ pound Swiss cheese, cut in ¼-inch cubes

1. Prepare the vegetables and boil water in a kettle. Preheat the oven to 375 degrees F.
2. Place 2 inches of boiling water in a 4-quart kettle. Arrange the peppers in a standing position and parboil for 6–8 minutes.
3. Drain the peppers; place them standing upright in a lightly greased baking dish. Combine the tomato and seasonings and spoon into the peppers, dividing the filling evenly. Mix the cheeses together and stuff into the peppers, rounding off the tops.
4. Bake until the peppers are hot and the cheese is melted, about 20 minutes.

SPINACH SURPRISE

6–8 servings
1 hour

1 cup unbleached flour
¾ teaspoon salt
1¼ teaspoons baking powder
2 tablespoons shortening, melted and cooled
¼ cup milk
4 eggs
¼ cup fresh parsley, finely chopped
½ cup diced onion
1½ cups cooked spinach, well drained
¼ cup grated Parmesan cheese
4 ounces Monterey Jack cheese, cut in ¼-inch cubes
1½ cups cottage cheese
½ teaspoon salt
2 cloves garlic, minced finely

1. Preheat the oven to 375 degrees F.
2. Grease a 12″ × 7½″ × 2″ baking dish.
3. Mix the flour, ¼ teaspoon of the salt, baking powder, shortening, milk, 2 of the eggs, parsley, and onion. Beat vigorously for 20 strokes. Spread the mixture in the baking dish. Combine the remaining ingredients, mix well, and spoon evenly over batter mixture.
4. Bake until set, approximately 30 minutes.
5. Let stand for 5 minutes before serving.

HARVEST-TIME ACORN SQUASH

4 servings
1¼ hours

2 medium acorn squash
boiling water
3 tablespoons melted butter
¼ cup brown sugar, packed
¼ teaspoon cinnamon
¼ teaspoon salt
⅛ teaspoon nutmeg
1 cup prepared applesauce
2 teaspoons butter

1. Preheat the oven to 375 degrees F. Begin boiling water in a kettle.
2. Wash the squash, cut in half lengthwise, remove seeds and stringy fibers. Place the squash, cut side down, on a rack in a shallow baking pan. Pour in ½ inch boiling water. Cover the pan with a tight-fitting lid or with tinfoil.
3. Bake for 30 minutes.
4. Remove the pan from the oven. Pour off the water, turn the squash cut side up. Combine the melted butter, 2 tablespoons of sugar, and the spices. Smooth over the edges of the squash pieces. Fill the centers with applesauce; top with the remaining brown sugar and butter.
5. Bake until the squash is tender, approximately 15 minutes.

CUCUMBERS IN DILL SAUCE

4 servings
35 minutes

4 slender 5-inch to 6-inch cucumbers
½ cup plain yogurt
½ cup sour cream
1 tablespoon chopped fresh dill or 1 teaspoon dried dill seed
¼ cup chopped green onions

1. Peel the cucumbers and slice thinly.
2. Combine the remaining ingredients and pour over cucumbers. Mix well.
3. Cover and chill in the freezer for ½ hour.

PASTA WITH GARDEN SALAD SAUCE

4 servings
½ hour

This is delicious served with Herb-Fried Chicken.

2 large tomatoes, scalded, peeled, and chopped
1 medium green pepper, chopped coarsely
1½ cups coarsely chopped broccoli
1 cup coarsely chopped onion
1 large clove garlic, minced finely
3 quarts water
¼ cup salad oil
3 tablespoons cider vinegar
½ teaspoon dried basil or 2 tablespoons fresh basil, chopped finely
4 cups precooked pasta (see p. 123)

1. Prepare the vegetables.
2. Put 3 quarts of water on to boil. Heat the salad oil in a large heavy skillet. Add all vegetables, *except the tomatoes*. Sauté the vegetables until tender crisp.
3. Add the tomatoes, vinegar, basil. Cook for 2 minutes. Keep warm.
4. Drop the pasta into the boiling water. Return the water to a boil. Turn off the heat and let the pasta remain in the water for 1 minute.
5. Drain the pasta. Top with vegetable sauce and serve.

SQUASH BISQUE

6–8 servings
½ hour

1½ cups diced onion
1 cup diced celery
1 cup chopped mushrooms
3 tablespoons chopped fresh parsley or 1 tablespoon dried
¼ cup chopped green pepper
¼ cup chopped sweet red pepper
2 cups mashed cooked winter squash
2 tablespoons butter
2 cups chicken stock, bouillon, or water
dash of cayenne pepper
1 cup milk
¼ cup light cream
salt and pepper to taste

1. Prepare the vegetables.
2. Melt the butter in a heavy 4-quart saucepan. Sauté the onion, celery, mushrooms, parsley, and green and red peppers until the onions are limp but not brown, about 5 minutes.
3. Add the squash, stock, cayenne pepper, and salt and pepper to taste. Simmer for 15 minutes.
4. Add the milk and cream, and reheat just to boiling.

DUTCH-STYLE GREEN BEANS

4 servings
15 minutes

3 slices bacon, diced
1 medium onion, diced
2 teaspoons cornstarch
¼ teaspoon salt
¼ teaspoon dry mustard
½ cup chicken broth or water
1 tablespoon cider vinegar
1 tablespoon brown sugar
2 cups chopped cooked green beans
1 hard-boiled egg, chopped

1. Sauté the bacon in a large skillet until crisp. Remove from the pan. Add the onion to the pan drippings and sauté until tender.

2. Blend in the cornstarch, salt, and mustard. Stir in the broth and cook, stirring constantly, until thickened.

3. Stir in the vinegar and sugar; add the beans and heat until bubbly. Serve with bacon and chopped egg.

LETTUCE AND GREEN PEA SOUP

4–6 servings
40 minutes

2 cups fresh peas (frozen peas may be substituted)
8 cups shredded crisp head lettuce
½ cup sliced green onions
1 clove garlic, minced finely (optional)
2 cups chicken broth or bouillon
1 teaspoon sugar
1 teaspoon salt
½ teaspoon dried chervil
dash pepper
1½ cups light cream

1. Shell the peas and chop the vegetables in a food processor.

2. In a heavy 4-quart Dutch oven, combine the lettuce, peas, onions, garlic, and broth. Cover the kettle and bring the contents to a boil. Reduce the heat and simmer for 15 minutes.

3. Pour half of the mixture into a blender and puree until smooth. Pour into a bowl. Put the balance of lettuce mixture in the blender with the sugar, salt, chervil, and pepper. Puree until smooth.

4. Return all the lettuce mixture to a saucepan if you are going to serve it warm and mix well, or mix well in a bowl to serve cold. Add the cream. Reheat or chill. Serve.

15 CREAMED VEGETABLE SOUPS

Each soup makes 4 servings

Many delicious fresh summer soups can be made from vegetable purees. These soups can be served hot or cold, as appetizers, or as the main meal. Either way they are quick and satisfying. They take just 10–15 minutes to make with prepared purees made from fresh cooked, canned, or frozen cooked vegetables. To make the puree, combine 3 cups of cooked vegetables with 1 cup of stock or water and puree in the blender. Purees for soup are best made with chicken stock, but vegetarians may wish to use water and increase the seasonings for extra flavor.

To make the soup, sauté the onion in butter until tender but not brown. Place in the blender with ½ of the puree and ½ of the cream or milk. Blend until smooth. Pour into a saucepan (or bowl, if to be eaten chilled). Blend the remaining ingredients; add to first half. Stir well. To serve warm, reheat just until piping hot. To serve cold, chill for several hours.

As with all soups, these are better if made ahead so that flavors can blend. All the soups can be garnished with chopped chives, parsley, croutons, bacon bits, yogurt, sour cream, or grated Parmesan cheese.

3 Cups Vegetable Puree	Onion	Butter	Light Cream or Milk	Seasonings
Beans, Green or Yellow	¼ cup	1 tablespoon	1 cup	2 dashes Tabasco, ¼ teaspoon garlic powder, ½ teaspoon basil, salt and pepper to taste
Beets	½ cup	1 tablespoon		¼ teaspoon garlic powder, 1 cup stock, 1 tablespoon lemon juice, ½ teaspoon celery salt, salt and pepper to taste, top with yogurt or sour cream
Broccoli	½ cup	2 tablespoons	1 cup	4 ounces cheddar cheese, salt and pepper to taste
Cabbage	¼ cup	1 tablespoon	1 cup	½ teaspoon celery salt, salt and pepper to taste
Carrots	¼ cup	2 tablespoons	1 cup	¼ teaspoon curry powder, salt and pepper to taste
Corn	⅛ cup	2 tablespoons	1 cup	1 cup diced chicken, 1 tablespoon chopped chives, salt and pepper to taste
Cucumbers	½ cup	2 tablespoons	1 cup	salt and pepper to taste, top with yogurt or sour cream
Greens	¼ cup	2 tablespoons	1 cup	½ teaspoon cayenne, salt and pepper to taste, top with yogurt or sour cream
Onions		2 tablespoons	1 cup	4 ounces cheddar cheese, 1 tablespoon cooking sherry, salt and pepper to taste
Peas, Green or Sugar Snap	½ cup	2 tablespoons	1 cup	½ teaspoon tarragon, salt and pepper to taste
Summer Squash	1 cup	2 tablespoons	1 cup	½ teaspoon curry powder, salt and pepper to taste
Tomatoes	½ cup	2 tablespoons		6 ounces American cheese, salt and pepper to taste
Winter Squash I	¼ cup	2 tablespoons	1 cup	½ teaspoon cayenne, salt and pepper to taste
Winter Squash II		2 tablespoons	1 cup	2 tablespoons brown sugar, ½ teaspoon allspice, ½ cup applesauce

Meatballs, meat patties, fried chicken, and similar foods can be tray frozen and packed in plastic bags. At mealtimes, take out the number of servings needed.

More Freeze-Ahead Ideas

Grate cheese for casseroles, pizza, and toppings and store in your freezer.

Prepare bread and cracker crumbs for toppings and coatings and store in your freezer.

Cook a large turkey with dressing. Slice the meat and cover with dressing and gravy. Freeze in meal-size portions.

Fix favorite foods or leftovers in individual foil or boilable-bag servings. Always chill any foods that have been precooked before packaging. Refrigerate for use within a day or so, or freeze for longer storage. Hurry-up meals can be prepared by just popping foil packages in the oven or boilable bags in a pan of hot water. There is little or no clean-up after.

Make extra pancakes and waffles and tuck some away in the freezer to give you a few extra minutes on summer mornings to package tray-frozen foods. The kids can fix their own breakfast by warming the waffles and pancakes in the toaster and topping them with fresh fruit, preserves, and a little yogurt.

Cook several pounds of ground beef; divide it into several containers to be tossed quickly into sauces and quick casserole dishes that require cooked ground beef.

If your family loves desserts, spring is the best time to freeze a few extra fruit breads, whole grain and fruit bars, cookies, and sheet cakes for busier times. Bake sheet cakes and divide in 4. Do not frost. These can be used later as single-layer cakes, double-layer birthday cakes (2 sections), or served with fresh fruit and toppings.

7 More Harvest-time Meal Planning Tips

Layer up a make-ahead salad in a large bowl. Layer shredded lettuce, peas, green onions, cooked macaroni, chopped tomatoes, sliced cucumbers, chopped walnuts, and diced chicken. Make a creamy dressing of 1½ cups mayonnaise, ¼ cup light cream, and 2 tablespoons chopped chives. Pour the

Foil line casserole dishes when making casseroles to be frozen. After the casseroles are frozen, they can be removed from the baking dish, freeing it for another use.

dressing over the salad but do not toss. Cover with foil or plastic wrap and chill for several hours or overnight.

Slow-cooked meals are time and energy savers. To adapt your own favorite stew recipe to the slow cooker, triple the recommended cooking time for the low setting or double it for the high setting. When cooking vegetables in the slow cooker, parboil them 6–10 minutes before adding to the pot. Always place vegetables in the bottom of the slow cooker with the meat on the top. Reduce the amount of liquid in your stew or casserole by about ⅓ and reduce the seasoning by at least that much.

Precook several servings of potatoes, rice, pasta, and hard-boiled eggs when you have an extra hour or so one day a week. They can be used for quick dishes during the week, such as potato salad, hash-browned potatoes, corned beef hash, creamed potatoes, soups, stir-fried rice, spaghetti, and egg salad.

Undercook pasta and rice by 4–5 minutes for baked dishes and 2–3 minutes for meals not requiring additional baking time. To reheat pasta for use with a quick sauce, bring lightly salted water to a boil, add pasta and return to a boil, drain, and add sauce. To reheat rice, place in a fine mesh sieve over boiling water, keep water simmering until rice is heated through, fluffing with a fork once or twice while heating; this just takes a few minutes.

Cook a large roast, meat loaf, turkey, pan of lasagna, or stew on the Friday evening of a harvest weekend. It will serve as a quick main meal with the addition of a fresh salad, bread, and beverage—even if company drops in. Leftover meats and meat loaf can be used for sandwiches.

Dice onions, celery, green peppers, and carrots. Place in plastic bags and refrigerate. This saves time when preparing soups, salads, and casseroles. If all the vegetables are not used up in 4–5 days, freeze the remainder; these will not be good for fresh salads after they are frozen, but can be used in all other ways including in stir-fried dishes and fried rice.

Tear salad greens (cut greens will turn brown around the edges when stored), wash, and spin dry; place them in plastic bags. They will maintain the crispness up to a week. To keep tomatoes firm when sliced ahead, slice them vertically and store in a separate bag.

Handy Reference Charts

Approximate Yield of Fruits From Fresh

Fruit	Fresh, as purchased or picked	Frozen Pints	Canned Quarts
Applesauce	1¼–1½ lb.	1	
Berries	1 crate (24 qt.)	32–36	12–18
	1⅓–1½ pt.	1	
Cantaloupes	1 dozen (28 lb.)	22	
	1–1¼ lb.	1	
Cherries, Sweet or Sour	1 bu. (56 lb.)	36–44	22–32
	1¼–1½ lb.	1	
Cranberries	1 box (25 lb.)	50	
	½ lb.	1	
Currants	2 qt. (3 lb.)	4	
	¾ lb.	1	
Peaches	1 bu. (48 lb.)	32–48	18–24
	1–1½ lb.	1	
Pears	1 bu. (50 lb.)	40–50	20–25
	1–1½ lb.	1	
Plums and Prunes	1 bu. (56 lb.)	38–56	24–30
	1–1½ lb.	1	
Raspberries	1 crate (24 pt.)	24	12–18
	1 pt.	1	
Rhubarb	15 lb.	15–22	
	⅔–1 lb.	1	
Strawberries	1 crate (24 qt.)	38	
	⅔ qt.	1	

Approximate Yield of Vegetables From Fresh

Vegetable	Fresh, as purchased or picked	Frozen Pints	Canned Quarts
Asparagus	1 crate (12 2-lb. bunches)	15–22	
	1–1½ lb.	1	
Beans, Lima (in pods)	1 bu. (32 lb.)	12–16	6–10
	2–2½ lb.	1	
Beans, Snap, Green, & Wax	1 bu. (30 lb.)	30–45	12–20
	⅔–1 lb.	1	
Beet Greens	15 lb.	10–15	3–8
	1–1½ lb.	1	
Beets (without tops)	1 bu. (52 lb.)	35–42	15–24
	1¼–1½ lb.	1	
Broccoli	1 crate (25 lb.)	24	
	1 lb.	1	
Carrots (without tops)	1 bu. (50 lb.)	32–40	16–25
	1¼–1½ lb.	1	
Cauliflower	2 medium heads	3	
	1⅓ lb.	1	
Chard	1 bu. (12 lb.)	8–12	3–8
	1–1½ lb.	1	
Collards	1 bu. (12 lb.)	8–12	3–8
	1–1½ lb.	1	
Corn, Sweet	1 bu. (35 lb.) (in husks)	14–17 (whole kernel)	6–10 (whole kernel)
	2–2½ lb.	1	
Kale	1 bu. (18 lb.)	12–18	3–8
	1–1½ lb.	1	
Mustard Greens	1 bu. (12 lb.)	8–12	3–8
	1–1½ lb.	1	
Peas (in pods)	1 bu. (30 lb.)	12–15	5–10
	2–2½ lb.	1	
Peppers, Sweet	⅔ lb. (3 peppers)	1	
Pumpkins	3 lb.	2	1
Spinach	1 bu. (18 lb.)	12–18	3–8
	1–1½ lb.	1	
Squash, Summer	1 bu. (40 lb.)	32–40	10–20
	1–1¼ lb.	1	
Squash, Winter	3 lb.	2	

Timetable for Blanching Vegetables for the Freezer

Vegetable	Blanch in Boilable Bag, 4 Bags at a Time (in minutes)	Blanch by Immersion or Steam, 1 Pound at a Time (in minutes)	Suitable for Freezing Unblanched
Beans, Green & Wax	6–8	3–4	Yes
Beets			No
Broccoli		5	Yes
Cabbage			Yes
Carrots	8–10	3–4	No
Corn on the Cob	10	6–8	Yes
Corn, Whole Kernel	6		No
Greens	Stir-fry until wilted: 2–3 minutes		No
Peas, Green	4	2	No
Summer Squash	5		Yes

Canning Times

Vegetables	Pressure Canning Time (in minutes)		Steam or Boiling Water Bath (in minutes)	
	pints	quarts	pints	quarts
Beans, Green & Wax	20	25		
Beets	30	40		
Carrots	25	30		
Corn, Whole Kernel	55			
Corn, Cream-style	85			
Greens	70	90		
Peas, Green	40	40		
Summer Squash	30	40		
Tomatoes, Puree			10	15
Cold Pack, Whole			35	45
Winter Squash & Pumpkin, Puree	65	85		

Adjusting for Altitude

Pressure Canner with a Dial Gauge

Altitude (in feet)	Process (in pounds)
0–2,000	10
2,000–3,000	11½
3,000–4,000	12
4,000–5,000	12½
5,000–6,000	13
6,000–7,000	13½
7,000–8,000	14
8,000–9,000	14½
9,000–10,000	15

If a pressure canner with a weighted gauge or control is used in a high altitude area, 2,000 feet or over, use 15 pounds pressure rather than 10 pounds pressure.

Boiling Water Bath and Steam Canner

Altitude (in feet)	If processing time called for is 20 minutes or less, increase by:	If processing time called for is more than 20 minutes, increase by:
1,000	1 minute	2 minutes
2,000	2 "	4 "
3,000	3 "	6 "
4,000	4 "	8 "
5,000	5 "	10 "
6,000	6 "	12 "
7,000	7 "	14 "
8,000	8 "	16 "
9,000	9 "	18 "
10,000	10 "	20 "

Suggested Reading

Ball Blue Book: The Guide to Home Canning and Freezing. Muncie, IN: Ball Manufacturing Co., 1979.

Down-to-Earth Vegetable Gardening Know-How featuring Dick Raymond. Charlotte, VT: Garden Way Publishing, 1975.

The Easy Harvest Sauce & Puree Cookbook by Marjorie Blanchard. Charlotte, VT: Garden Way Publishing, 1980.

The Food Preserver by the Editors of *Consumer Guide*. Skokie, IL: Publications International LTD., 1976.

Garden Way's Guide to Food Drying by Phyllis Hobson. Charlotte, VT: Garden Way Publishing, 1980.

Garden Way's Zucchini Cookbook by Nancy Ralston and Marynor Jordan. Charlotte, VT: Garden Way Publishing, 1977.

Home Gardening Wisdom by Dick and Jan Raymond. Charlotte, VT: Garden Way Publishing, 1982.

How to Live on Almost Nothing and Have Plenty by Janet Chadwick. New York: Knopf, 1979.

Keeping the Harvest by Nancy Chioffi and Gretchen Mead. Charlotte, VT: Garden Way Publishing, 1980.

Kerr Home Canning and Freezing Book. Sand Springs, OK: Kerr Manufacturing Co.

Putting Foods By by Ruth Hertzberg, Beatrice Vaughan, and Janet Greene. Brattleboro, VT: Stephen Greene Press, 1975.

Index